Water Interfaces in Physics, Chemistry and Biology: A Multi-Disciplinary Approach 2007

Journal of Physics: Conference Series Volume 177

Innsbruck, Austria
8-13 December 2007

ISBN: 978-1-61738-306-9

Printed from e-media with permission by:

Curran Associates, Inc.
57 Morehouse Lane
Red Hook, NY 12571
www.proceedings.com

Some format issues inherent in the e-media version may also appear in this print version.

Printed by Curran Associates, Inc. (2010)

Institute of Physics
Dirac House, Temple Back
Bristol BS1 6BE UK

Phone: 44 1 17 929 7481
Fax: 44 1 17 920 0979

techtracking@iop.org

TABLE OF CONTENTS

PAPERS

EXTENDED ABSTRACTS

Journal of Physics: Conference Series **177** (2009) 011001 doi:10.1088/1742-6596/177/1/011001

ESF-conference: Obergurgl, 8-13 December 2007

Title: Water Interfaces in Physics Chemistry and Biology: a multi-disciplinary approach

This 5-day meeting, sponsored by the European Science Foundation (ESF) in partnership with the Fonds zur Förderung der wissenschaftlichen Forschung in Österreich (FWF) was organised by Marie-Claire Bellissent-Funel *[Lab. Léon Brillouin (CEA-CNRS), CEA Saclay]* and John Dore *[School of Physical Sciences, University of Kent]*. It took place in the Universitatszcentrum *[University of Innsbruck]*, in the ski resort of Obergurgl, Austria, during 8th-13th Dec, 2007.

The main aim of the meeting was to bring together various groups working on the characteristics of water in a wide range of different conditions, particularly in relation to the difference in behaviour for bulk water and water in the close proximity to an interface. Another focus was on the properties of 'solid water' and the free time during the afternoon provided a good opportunity for studying ice interfaces in a different context as the snow conditions were good for ski-ing !

An outline of the programme is indicated below:-

Session 1: Water/ice characteristics

(i) Structure and dynamics, hydrogen-bonding, water-ice transformations
(ii) Spectroscopic probes: neutrons, X-rays, NMR, FTIR, Raman

Mark E. Tuckerman
New York University, US
How accurately can generalized-gradient approximations to density functional theory describe the properties of water and aqueous solutions? Results from ab initio molecular dynamics performed in the complete basis set limit

José Teixeira
LLB CEA-CNRS, CEA Saclay, FR
Study of the dynamics of hydrogen bonds in water and consequences for the unusual behaviour of supercooled water

Helmut Schober
Institut Laue Langevin, Grenoble, FR *via the microscopic dynamics and via the kinetics of the transformation processes*
Characterization of amorphous ice structures

Werner Kuhs
University of Göttingen, DE
Gas - water interactions in crystalline ices and gas hydrates

Sessions 2 and 3 : Water in confined geometry/ Solid Substrates

(i) Interfacial interactions : hydrophobicity/hydrophilicity
(ii) Surface characterisation
Silicas (sol-gel, Vycor, CPG, MCM, SBA……), Zeolites, clays
Activated carbons, carbon nanotubes.
Structure and dynamics (constrained molecular mobility), water-ice transformations

John Dore
University of Kent, Canterbury, UK
Structural studies of water in a hydrophobic environment

Katsumi Kaneko
Chiba University, JP
Structure of water and hydrated ions confined in carbon nanospaces

Gerhard Findenegg
Technische Universität Berlin, DE
Limits of freezing and melting of water in MCM-41type cylindrical nanopores

Sow-Hsin Chen
MIT, Cambridge, US
Experimental observation of fragile-to-strong dynamic crossover in confined and hydration water and its relation to the liquid-liquid critical point in supercooled water

Harald Reichert
Max-Planck-Institut für Metallforschung, Stuttgart, DE
X-ray investigations of the structure of water and ice at interfaces

Harald Reichert
Max-Planck-Institut für Metallforschung, Stuttgart, DE
X-ray investigations of the structure of water and ice at interfaces

Journal of Physics: Conference Series **177** (2009) 011001 doi:10.1088/1742-6596/177/1/011001

Volker Kempter
Technische Universität Clausthal, DE
Molecules at the interface between vacuum and solid water

Harald Reichert
Max-Planck-Institut für Metallforschung, Stuttgart, DE
X-ray investigations of the structure of water and ice at interfaces

Sessions 4 and 5: Water in confined geometry/ Soft Systems

(i) Related systems-clathrates, aqueous solutions, amphiphiles, surfactants, lamellar liquid crystals, microemulsions, soluble polymers
(iii) Structure and dynamics (constrained molecular mobility), hydrogen-bonding
(iii) Spectroscopic probes: neutrons, X-rays, NMR, FTIR, Raman

Volker Kempter
Technische Universität Clausthal, DE
Molecules at the interface between vacuum and solid water

Imre Bako
Hungarian Academy of Sciences, Budapest, HU
Investigation of the structure of aqueous mixtures and electrolyte solutions by different methods

Harald Reichert
Max-Planck-Institut für Metallforschung, Stuttgart, DE
X-ray investigations of the structure of water and ice at interfaces

Volker Kempter
Technische Universität Clausthal, DE
Molecules at the interface between vacuum and solid water

Imre Bako
Hungarian Academy of Sciences, Budapest, HU
Investigation of the structure of aqueous mixtures and electrolyte solutions by different methods

John Finney
University College London, UK
Solvent effects in the formation and stabilisation of molecular structures and assemblies: the roles of order and disorder

Jan Swenson
Chalmers University of Technology, Göteborg, SE
Dynamics of hydration water and its implication for biomolecular dynamics

Session 6: Water in confined geometry/ Bio-Systems

(i) Biological interfaces, membranes, proteins, biological complexes, cells
(ii) Interfacial water in biological processes, enzymatic catalysis, water channels, ionic channels
(iv) Spectroscopic probes: neutrons, X-rays, NMR, FTIR, Raman

Bertil Halle

Lund University, SE
Exploring biomolecule-water interaction by NMR

Joseph Parello
Vanderbilt University, Nashville, US
Hydration in GPCR-mediated signal transduction

Giuseppe Zaccai
Institut Laue Langevin, Grenoble, FR
Water in halophile Dead Sea organisms: neutron scattering studies of adaptation to extreme environments

Sessions 7 and 8 : Water in confined geometry/ Bio-Systems/Modelling

(i) Molecular modelling, molecular dynamics, DFT computations, etc
(ii) Interaction potentials, simple and otherwise, ab initio methods
(iii) Hydrogen-bond effects on structure and dynamics in confinement

Branka M. Ladanyi
Colorado State University, Fort Collins, US
Structural order and dynamics at water/hydrocarbon interfaces and in reverse micelles

Bert de Groot
MPI for Biophysical Chemistry, Göttingen, DE
Dynamics and energetics of water permeation across biological channels and model peptidic pores

Pavel Jungwirth
Academy of Sciences of the Czech Republic, Prague, CZ
Ions at surfaces of hydrated proteins and other interfaces

Poster Sessions, round table discussions

Participants who receive ESF support will be expected to display a poster of their work at the symposium and to engage in discussions with viewers during the period set aside for the poster viewing. There will also be an allocated session for poster discussion, including some oral presentations.

One panel discussion has been held to address any points of particular controversy and a four-member panel will be primarily drawn from the invited lecturers. It is impossible to specify what topics might be of topical concern at the time of the meeting but it is likely that suitable issues will eventually present themselves and can be adapted to the circumstances.

The symposium will end with a general debate about important current issues and possible future developments, following the recommended form of a 'Forward Look Plenary Discussion'.

Journal of Physics: Conference Series **177** (2009) 011001 doi:10.1088/1742-6596/177/1/011001

The presentations and the complete programme are on the conference website :

www.esf.org/conferences/07225

There was a wide representation encompassing 30 countries and 130 scientists drawn from different science disciplines. Furthermore there was a good range of young scientists, who made an excellent contribution to the poster session. There were, of course, many animated discussions away from the conference room and the feedback forms showed that almost everyone [96% !] felt that they had enjoyed the sessions and had learned something new. There was support for a further conference on this theme in the future.

Inevitably, many of the speakers presented information that was in preparation for publication elsewhere and therefore our compilation of some papers in this brief report is not fully representative of the range of topics discussed at the meeting. Further information on specific work reported at the meeting can be obtained by following the author list through the Web of Science or by contacting the authors directly. We report eight short papers from the meeting, as well as five extended abstracts:

1. Initial Filling Mechanism of Predominant Water Adsorption on Hydrophobic Slit-Shaped Carbon Nanopores
Tomonori Ohba and Katsumi Kaneko

2. Computer simulation study of water/hydrocarbon interfaces: effects of hydrocarbon branching on interfacial properties
Janamejaya Chowdhary and Branka M. Ladanyi

3. Thermodynamics of supercooled water in solutions
D. Corradini, P. Gallo and M. Rovere

4. Transferability of polarizable models for ion-water electrostatic interaction
Marco Masia

5. Quantum chemical study of water impact on the calcium hydroxyapatite
V. D. Khavryuchenko, O.V. Khavryuchenko, V.V. Lisnyak

6. Neutron Scattering Studies of Dynamic Crossover Phenomena in a Coupled System of Biopolymer and Its Hydration Water

Sow-Hsin Chen[1], F. Mallamace[1, 2], X. Q. Chu[1], C. Kim[1], M. Lagi[1, 4], A. Faraone[3], E. Fratini[4], P. Baglioni[4]

7. Looking for the best experimental conditions to detail the protein solvation shell in a binary aqueous solvent via Small Angle Scattering
Maria Grazia Ortore, Raffaele Sinibaldi, Francesco Spinozzi, Andrea Carbini, Flavio Carsughi and Paolo Mariani

8. Influence of the water molecules near surface of viral protein on virus activation process
S. O Shepelenko, A. S. Salnikov, S. V. Rak, E. P. Goncharova and A. B. Ryzhikov

9. Optical Kerr effect measurements on supercooled water: the experimental perspectives
P. Bartolini, A.Taschin, R. Eramo, R. Righini and R.Torre

10. Structural studies of water confined in a confined hydrophobic environment
J.C. Dore, M.-C. Bellissent-Funel, A. Burian, H.P. Castricum, J. Jelassi, K. Kaneko, T. Ohba, H. Tanaka and J.B.W. Webber

11. Dynamic transition and glassy behaviour in hydrated proteins
F.Mezei, M. Russina, G. Chen, H. Frauenfelder, P. W. Fenimore, P. Falus and B. Farago

12. Relation between frequency and H bond length in heavy water: Towards the understanding of the unusual properties of H bond dynamics in nanoporous media
S. Pommeret, R. Musat, J.- P. Renault, J.-C. Leicknam and S. Bratos

13. Quantum confinement of hydrogen in ice based clathrates
M. Russina, E. Kemner, M. Celli, L. Ulivi and F.Mezei

Marie-Claire Bellissent-Funel [marie-claire.bellissent-funel@cea.fr]
John Dore [j.c.dore@kent.ac.uk]

All papers published in this volume of *Journal of Physics: Conference Series* have been peer reviewed through processes administered by the proceedings Editors. Reviews were conducted by expert referees to the professional and scientific standards expected of a proceedings journal published by IOP Publishing.

Initial Filling Mechanism of Predominant Water Adsorption on Hydrophobic Slit-Shaped Carbon Nanopores

Tomonori Ohba and Katsumi Kaneko

Graduate School of Science, Chiba University, 1-33 Yayoi, Inage, Chiba 263-8522, Japan

E-mail: ohba@pchem2.s.chiba-u.ac.jp, kaneko@pchem2.s.chiba-u.ac.jp

Abstract. Predominant water adsorption in hydrophobic slit-shaped carbon nanopores begins at a threshold pressure of water vapor. Adsorption of water vapor was faster for pores of 0.7 nm in width than for pores of 1.1 nm in width. The formation of water dimers is the rate-determining step of the water adsorption from analysis of the adsorption kinetics. The adsorption amount from grand canonical Monte Carlo simulation oscillated with the calculation cycle up to an adsorption amount of 30 mg ml^{-1} and then steadily increased to an equilibrium adsorption amount. The Gibbs free energy change with the cluster size of water molecules has a barrier for the cluster growth below the critical adsorption density. The snapshot analysis of the grand canonical Monte Carlo simulation showed that formation of nanoscale clusters of water molecules begins just at the critical density of 30 mg ml^{-1}. Then the nanoscale cluster formation should be essential to induce the predominant adsorption at the threshold pressure corresponding to the critical adsorption density.

Introduction

Water adsorption on nanoporous carbons has been actively studied for a long time [1-5]. As the water adsorption isotherm varies sensitively with the change of surface functional groups, the relationship between the water adsorption isotherm and surface functional groups has been studied using nanoporous carbons having different surface functional groups [6-10]. Water molecules are adsorbed on the surface functional groups with hydrogen bonding even below the relative pressure (P/P_0) of 0.05; these nanoporous carbons having surface functional groups are essentially hydrophilic. Phenomenologically, water adsorption on the nanoporous carbons of hydrophilicity can be briefly described by so-called Dubinin-Serpinsky equation [11]. Talu and Meunier introduced a structural view to the water adsorption mechanism [12]. However, nanoporous carbons of only slight amount of surface functional groups exhibit the water adsorption isotherm completely different from that of hydrophilic carbons; the water adsorption below $P/P_0 = 0.3$ is almost negligible and predominant adsorption begins above $P/P_0 = 0.3$ [13]. Hence, these nanoporous carbons should be originally hydrophobic. Pitch-based activated carbon fiber (P-ACF) and single wall carbon nanotubilities of high purity are representatives of hydrophobic nanoporous carbons. Even these hydrophobic nanoporous carbons have slight amount of surface functional groups which can be detected by X-ray photoelectron spectroscopy, the water adsorption isotherm does not change so much by reduction treatment of surface functional groups with Ar or H_2 at high temperature [14]. Phenomenologically a clear hydrophobicity can be confirmed by formation of a hemispherical water droplet on these carbon pellets, as shown in Fig. 1. Accordingly, we can name these nanoporous carbons hydrophobic nanoporous carbons. In particular, P-ACF is a carbon sample fit for study on water adsorption on the hydrophobic nanoporous carbon according to authors' systematic characterization using various techniques [15-19]. P-ACFs have considerably uniform slit-shaped nanopores of which pore width ranges from 0.6 to 1.4 nm. Then, the P-ACFs have been used for fundamental researches on confinement effect for molecules and ions.

Authors have shown that the structure of water and CCl_4 adsorbed in the nanopores depends on the pore width using *in situ* X-ray diffraction [20-22], and unique magnetism of O_2 adsorbed varies with the pore width [23-25]. In the case of water adsorption, the electron radial distribution function (ERDF) of water-adsorbed P-ACF (pore width w = 0.7 nm) at 303 K is almost similar to that at 170 K, whereas the ERDF of water-adsorbed P-ACF (w = 1.2 nm) depends sensitively on the measuring temperature; water in the 1.2 nm-pore at 303 K is still distinctly different from that of bulk liquid water, being rather close to a solid-like structure. The fact of solid-like structure of adsorbed water in carbon nanopores was ascertained with X-ray diffraction and neutron diffraction by Bellissent-Funel et al [26]. Thus, water molecules are predominantly adsorbed in hydrophobic nanopores of carbon, forming a highly stabilized solid-like structure. Then, there are fundamental questions why and how water molecules are adsorbed in the hydrophobic nanopores. As to the stabilization of adsorbed water, these authors showed that cluster formation of water molecules gave sufficient stabilization energy mainly stemming from hydrogen bonding among water molecules in the cluster [5]. Then, one of the above fundamental questions is resolved. However, still we wonder how water molecules are adsorbed above P/P_0 = 0.3. This article describes the study on this question. These are another essential question on water adsorption in hydrophobic carbon nanopores. The water adsorption isotherm has an explicit adsorption hysteresis. For example, Miyawaki et al. showed that the adsorption hysteresis loop depended on the pore width in the range of 0.7 to 1.4 nm; the adsorption hysteresis almost disappeared for the 0.7 nm-pores [27]. Similar results were often reported [28-30]. Iiyama et al. introduced *in situ* small angle X-ray scattering (SAXS) to obtain the density fluctuation in carbon nanopores having adsorbed water, although an exact correction of the water molecule-carbon wall scattering was difficult; they showed a clear difference of the density fluctuation for adsorption and desorption branches [31]. Ohba and Kaneko improved the *in situ* SAXS method to add the GCMC simulation; the GCMC-aided *in situ* SAXS suggested plausible distribution models of water molecules of water molecules and/or cluster in the carbon nanopores for adsorption and desorption branches [32,33]. Therefore, we can understand structurally both the adsorbed states of adsorption and desorption processes.

In this article, we describe the induction mechanism of the predominant water adsorption in the representative nanopores of p-ACFs at a critical pressure above P/P_0 = 0.3.

1. Simulation and Experimental Section

GCMC simulation was used to study on the water structure in the hydrophobic nanopores of w = 0.7 and 1.1 nm at 303 K, assuming a unit cell of $6\times6\times H$ nm^3, which was the same method as that in our earlier publication [34]. Here H was the physical pore width defined as the distance between carbon-atomic centers across the nanopore. In each calculation cycle, random selection among three events of displacement, insertion, and removal was chosen such that there were equal probabilities of insertion and removal. The number of the calculation cycle was 10^7 steps and the last 10^4 steps were taken for the statistics collection in each adsorption point. The arrival at equilibrium was checked with number change of molecules within 1% for 3×10^5 steps. The saturated vapor pressure P_0 was corrected by the pressure of bulk condensation in GCMC simulation. The calculation in canonical ensemble Monte Carlo simulation was done for 10^6 steps to determine accurate water distribution from the water distributions after the calculation of GCMC simulation with 10^7 steps in the calculation cycle.

The five-site transferable interaction potential (TIP5P) model of a water molecule was used for simulating water adsorption, which was combination of the Lennard-Jones interaction between oxygen atoms and the electrostatic interaction between partial charges on the site of hydrogen atoms and lone pairs [35].

$$\phi_{ij}(r) = 4\varepsilon_{ij}\left[\left(\frac{\sigma_{ij}}{r}\right)^{12} - \left(\frac{\sigma_{ij}}{r}\right)^{6}\right] + \sum_{i}^{4}\sum_{j}^{4}\frac{1}{4\pi\varepsilon_0}\frac{q_i q_j}{r_{ij}} \tag{1}$$

Here ε_{ij} and σ_{ij} are the water potential well depth (80.5 K) and the effective diameter (0.312 nm), respectively. The absolute value of the partial charge q_i is 3.86×10^{-20} C. The details of the model are given by Mahoney et al [31]. We assume that a carbon atom is neutral and then the interaction between water-graphitic carbon wall comes from the sum of the Lennard-Jones interaction. Thus, the interaction between a water molecule and graphitic carbon wall is represented by the potential function of the infinite carbon planes of graphite using Steele potential model [36],

$$\phi_{sf}(z) = 2\pi\sigma_{sf}^{2}\varepsilon_{sf}\rho\Delta_C\left[\frac{2}{5}\left(\frac{\sigma_{sf}}{z}\right)^{10} - \left(\frac{\sigma_{sf}}{z}\right)^{4} - \frac{\sigma_{sf}^{\ 4}}{3\Delta_C(z+0.61\Delta_C)^3}\right] \tag{2}$$

where σ_{sf} and ε_{sf} are the water-carbon atomic distance and effective diameter, fitted by using Lorentz-Berthelot rule (σ_{sf} = 0.327 nm and ε_{sf} / k_B = 49.3 K). The symbols of ρ, Δ_C, and z represent the carbon number density of graphite, the interlayer distance between graphite sheets, and the vertical distance of a water molecule from the atomic center of the carbon atom on surface, respectively. Then the water-carbon interaction in the nanopores is given by the sum of the interaction from both carbon walls [34].

$$\phi_p = \phi_{sf}(z) + \phi_{sf}(H-z),\ H = w_{eff} + 2z_0,\ z_0 = 0.856\sigma_{sf} \tag{3}$$

z_0 is the contact distance between a water molecule and a graphite wall. We relate the physical pore width H to the effective pore width w_{eff}, which is determined experimentally. Free energy change on cluster formation at an adsorbed amount was evaluated by the calculation of enthalpy and entropy terms using water cluster model in a unit cell. The enthalpy was determined by the adsorption energy obtained from the sum of the interaction potential. The entropy term was derived from the statistical state number calculated by simple site occupancy model for water clusters in a nanopore associated with degree of freedom in a cluster. We calculated site occupancy by associated molecules. Then the entropy was evaluated from the number of states from the site occupancy and freedom of associated molecules.

The pitch-based activated carbon fibers (P-ACFs) of average pore widths of 0.7nm (P-ACF0.7) and 1.1 nm (P-ACF1.1) were used as model slit pore systems for water adsorption at 303 K. Slight amount of surface oxygen was determined by X-ray photoelectron spectroscopy (XPS) (Mg Kα, 10 kV, 10 mA); the surface oxygen/carbon ratio was determined from the peak intensities of O1s and C1s XPS. The adsorption isotherms of both P-ACFs were volumetrically measured at 303 K after preheating in a vacuum below 1 mPa at 383 K for 2 hours. The equilibrium adsorption amount was determined when the gas phase pressure change become less than 10 Pa for an hour. The change in the amount of water adsorbed on P-ACFs was observed every 0.1 sec for evaluation of adsorption kinetics, when the initial change in the adsorbed amounts is up to 150 mg ml^{-1}, corresponding with the vapor pressure changes from vacuums to 1.9 kPa for P-ACF0.7 and 2.7 kPa for P-ACF1.1.

2. Results and Discussion

Figure 2 (a) shows the experimental water adsorption isotherms at 303 K on P-ACF0.7 and P-ACF1.1. A clear hysteresis loop is observed for P-ACF1.1, while P-ACF0.7 has almost no hysteresis loop, as previous reported [27, 37]. An adsorbed amount gradually increases with increasing pressure up to

P/P_0 = 0.3 and then steeply increases for P-ACF0.7. A steep increase in adsorbed amount occurs at P/P_0 = 0.7-0.8 for P-ACF1.1. The difference in the rising behavior of the adsorption isotherms for P-ACF0.7 and P-ACF1.1 is not caused by different concentration of the surface oxygen. The XPS examination showed that the O1s/C1s peak intensities for P-ACF0.7 and P-ACF1.1 were 0.08 and 0.07, respectively, being almost similar each other. The distinctly different adsorption behavior below P/P_0 = 0.5 for P-ACF0.7 and P-ACF1.1 must stem from the pore width difference. Hence, a gradual adsorption below P/P_0 does not indicate the presence of highly concentrated surface oxygen groups. The predominant adsorption is also observed in simulated adsorption isotherms of both nanopores, as shown in Figure 2 (b); the abrupt increases of adsorbed amount for the 0.7 and 1.1 nm-nanopores are observed at P/P_0 = 0.4 and 0.6, respectively. These threshold pressures of the predominant adsorption roughly correspond to these of experimental adsorption isotherms, although the ascending pressure of the simulated adsorption isotherm for the 1.1 nm-nanopore is close to the descending pressure of the experimental desorption isotherm. Hence the adsorption branch for the 1.1 nm-nanopores should stem from a metastable adsorption process. As the kinetic data can provide information on the elementary adsorption process on both ACFs, experimental adsorption rates of water are obtained, as shown in Figure 3. Water adsorption almost reaches each equilibrium after 100 sec for P-ACF0.7 and 300 sec for P-ACF1.1. Thus, P-ACF of narrower pore width induces faster adsorption of water vapor. The chemical engineering analysis of the water adsorption process with an *n*-th order reaction model was used for determination of the elementary adsorption process. The best fitting curve for the adsorption rate is obtained, when *n* is two. Here the second-order rate coefficients for P-ACF0.7 and P-ACF1.1 are 9.6×10^{-4} and 3.5×10^{-4} ml mg^{-1} sec^{-1}, respectively. The best fitting to the second order reaction (n=2) indicates that the rate-determining step is water dimer formation. Then bigger water clusters should be immediately formed after the dimer formation. Thus, the dimer formation is a key for the initial cluster growth in the predominant adsorption of water in the hydrophobic nanopores. The dimers should be more stabilized in the narrower pores [5] and thereby adsorption of water starts gradually through the dimer formation even below P/P_0= 0.3 only for P-ACF0.7. Still we have a question why adsorption begins just at the threshold pressure.

We simulated the adsorption amount changes as a function of a simulation step just above the threshold pressure (P/P_0 = 0.4 for the pore of w = 0.7 nm and P/P_0 = 0.6 for the pore of w = 1.1 nm), as shown in Figure 4 in order to figure out adsorption dynamics at the threshold pressures of adsorption from GCMC simulation. The simulated adsorption amount changes show the tendency similar to experimental adsorption kinetics. The increase of an adsorption amount is faster for the narrower nanopore; the beginning of the remarkable adsorption is observed around 2×10^6 simulation steps for the 0.7 nm-nanopore and 3×10^6 simulation steps for the 1.1 nm-nanopore. Then we can say that adsorption for the 0.7 nm-nanopore is faster than that for the 1.1 nm-nanopore. The simulation trend of the adsorption rate agrees with the above-mentioned experimental results. It is noteworthy that the simulated adsorption amounts with calculation cycle are scattered below 30 mg ml^{-1} of the adsorption amount for both pore systems, giving a stationary increase above it. The amount of 30 mg ml^{-1} should be a critical density for inducing the stationary adsorption of water in hydrophobic nanopores. We calculated the free energy change per a water molecule with the molecular association number to obtain the thermodynamic understanding, as shown in Fig. 5. The relationship between the free energy and molecular association number has a peak-top at dimer below 30 mg ml^{-1} of the adsorbed density. Then, isolated water molecules have to go over an energy barrier of the peak-top at the dimer for successive cluster growth below 30 mg ml^{-1} of the adsorbed density. The relationship having the peak-top transforms into the simple reduction of the free energy with the association number at 30 mg ml^{-1} of the adsorbed density. This means that the cluster growth proceeds without an energy barrier above 30 mg ml^{-1} of the adsorbed density. Then the water dimers spontaneously grow into nanoclusters such as pentamers and hexamers above 30 mg ml^{-1} of the adsorbed density. Thus, the free energy changes provide the reason for the water dimer formation as the rate-determining step and presence of the critical density of 30 mg ml^{-1}.

The spontaneous growth of water clusters beyond the critical density was confirmed by GCMC simulation. Figure 6 depicts snapshots of water structure in the nanopores of the w = 0.7 and 1.1 nm.

The isolated water molecules appear at 15 mg ml^{-1} of the adsorbed density. Small clusters such as dimers, trimers, and tetramers are formed at 30 mg ml^{-1} of the adsorbed density. Water clusters gather mainly along carbon walls and merge together to form pentamers, hexamers, and larger clusters beyond 30 mg ml^{-1} of the adsorbed density. Thus, water molecules form primary clusters of the necessary concentration at 30 mg ml^{-1} of the critical density. The number distribution of associated molecules in a cluster is obtained from the snapshots, which explains well the above discussion, as shown in Figure 7. A maximum of the association number distribution of clusters shifts from zero to 5-7 with increasing the adsorbed density from 30 mg ml^{-1} to 120 mg ml^{-1}. Thus, the adsorbed density of 30 mg ml^{-1} is a critical factor for the successive growth of clusters. These results agree with the free energy change in Fig. 5. Then, water molecules in the nanopore remain mainly as monomers, dimers, and trimers below the critical density. The change in an averaged cluster diameter with the adsorbed density can be evaluated from the number distribution of association molecules in Figure 7 under the assumption of a spherical cluster shape, as shown in Fig. 8. For both nanopores, the mean cluster diameter steeply increases up to 0.5 nm until 60 mg ml^{-1}of the adsorbed density, accompanying with an only slight increase. Thus, it is suggested the two-stage growth mechanism of clusters, that is, growth up to hexamer-order clusters below 60 mg ml^{-1} of the adsorbed density and the number increase of the hexamer-order clusters. The experimentally observed size of unit clusters in the preceding SAXS experiment [32] coincides well with the above results.

The present study shows that water adsorption in hydrophobic nanopores of w = 0.7 and 1.1 nm suddenly start above the critical adsorbed density of 30 mg ml^{-1}. The critical density gives 1.0 nm of the average intermolecular distance of water, almost corresponding to tri-molecular size of the water molecule; this intermolecular distance should be associated with the percolation threshold. The primary clusters such as dimers and trimers merge each other to stable unit clusters of hexamer-order size beyond the critical adsorbed density.

Acknowledgments

This work was supported by the Kurata Memorial Hitachi Science and Technology Foundation.

References

[1] McBain J W, Porter J L and Sessions R F 1933 *J. Am. Chem. Soc.* **55** 2294
[2] Pidgeon L M and Maass O 1930 *J. Am. Chem. Soc.* **52** 1053
[3] Rao K S 1941 *J. Phys Chem.* **45** 500
[4] Mowla, D. Do, D. D.and Kaneko, K. 2003, *Chem. Phys. Carbon,* **28**, 229.
[5] Ohba T, Kanoh H and Kaneko K 2004 *J. Am. Chem. Soc.* **126** 1560
[6] Salame I I and Bandosz T J 1999 *Langmuir* **15** 587
[7] Julien F, Baudu M and Mazet M 1998 *Carbon* **32** 3414
[8] Lee W H and Reucroft P J 1999 *Carbon* **37** 7
[9] Müller E A, Rull L F, Vega L F and Gubbins K E 1996 *J. Phys. Chem.* **100** 1189
[10] Jorge M, Schumacher C and Seaton N A 2002 *Langmuir* **18** 9296
[11] Dubinin M M and Serpinsky V V 1981 *Carbon* **19** 402
[12] Talu O and Meunier F 1996 *AIChE J.* **42** 809
[13] Kaneko K, Hanzawa Y, Iiyama T, Kanda T and Suzuki T 1999 *Adsorption* **5** 7
[14] Ohba T and Kaneko K 2007 *J. Phys Chem C* **111** 6207
[15] Setoyama N, Ruike M, Kasu T, Suzuki T and Kaneko K 1993 *Langmuir* **9** 2612
[16] Ishii C, Suzuki T, Shindo N and Kaneko K 1997 *J. Porous Mater.* **4** 181
[17] El-Merraoui M, Aoshima M and Kaneko K 2000 *Langmuir* **16** 4300
[18] Shin S, Jang J, Yoon, S –H and Mochida I 1997 *Carbon* **35** 1739
[19] Ohba T, Nicholson D and Kaneko K 2003 *Langmuir* **19** 5700

[20] Iiyama T, Nishikawa K, Suzuki T and Kaneko K 1997 *Chem. Phys. Lett.* **274** 152
[21] Iiyama T, Nishikawa K, Otowa T and Kaneko K 1995 *J. Phys. Chem.* **99** 10075
[22] Iiyama T, Nishikawa K, Suzuki T, Otowa T, Hijiriyama M, Nojima Y and Kaneko K 1997 *J. Phys. Chem. B* **101** 3037
[23] Kaneko K, Suzuki J and Ishii C 1998 *Chem. Phys. Lett.* **282** 176
[24] Kanoh H, Zamma A, Setoyama N, Hanzawa Y and Kaneko K 1997 *Langmuir* **13** 1047
[25] Kanoh H and Kaneko K 1996 *J. Phys. Chem.* **100** 755
[26] Bellissent-Funel M –C, Sridi-Dorbez R and Bosio L 1996 *J. Chem. Phys.* **104** 10023
[27] Miyawaki J, Kanda T and Kaneko K 2001 *Langmuir* **17** 664
[28] Branton P J, Hall P G and Sing K S W 1995 *Adsorption* **1** 77
[29] Lagorsse S, Campo M C, Magalhães F D and Mendes A 2005 *Carbon* **43** 2769
[30] Striolo A, Gubbins K E, Gruszkiewicz M S, Cole D R, Simonson J M, Chialvo A A, Cummings P T, Burchell T D and More K L 2005 *Langmuir* **21** 9457
[31] Iiyama T, Ruike M and Kaneko K 2000 *Chem Phys Lett* **331** 359
[32] Ohba T, Kanoh H and Kaneko K 2005 *Chem. Eur. J.* **11** 4890
[33] Ohba T, Kanoh H and Kaneko K 2005 *Nano Lett.* **5** 227
[34] Ohba T, Kanoh H and Kaneko K 2004 *J. Phys. Chem. B* **108** 14964
[35] Mahoney M W and Jorgensen W L 2000 *J. Chem. Phys.* **112** 8910
[36] Steele W A 1973 *Surf. Sci* **36** 317
[37] Ohba T and Kaneko K 2007 *Mol. Phys.* **105** 139

Figure 1. Image of a water droplet on a pellet of the P-ACF powder after reduction treatment in Ar atmosphere at high temperature at 0.1 seconds after contact of the water droplet with the pellet.

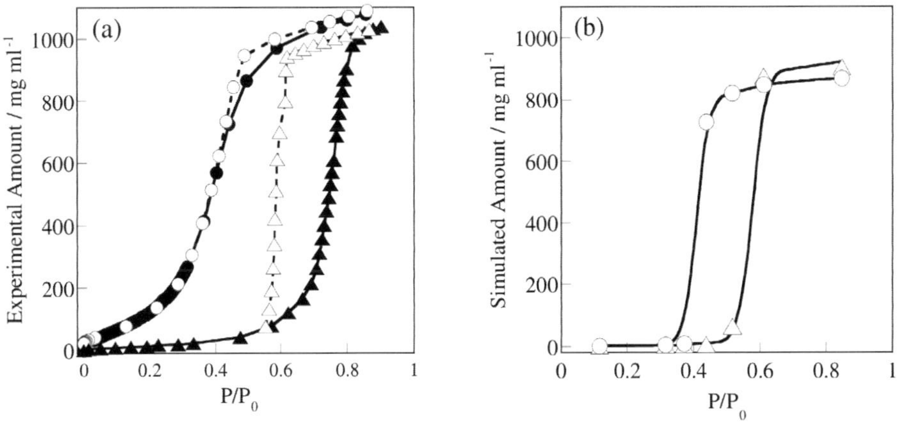

Figure 2. Water adsorption isotherms in experiment (a) and simulation (b). Circle: w = 0.7 nm and triangle: w = 1.1 nm. Closed symbol: Adsorption branch and opened symbol: Desorption branch.

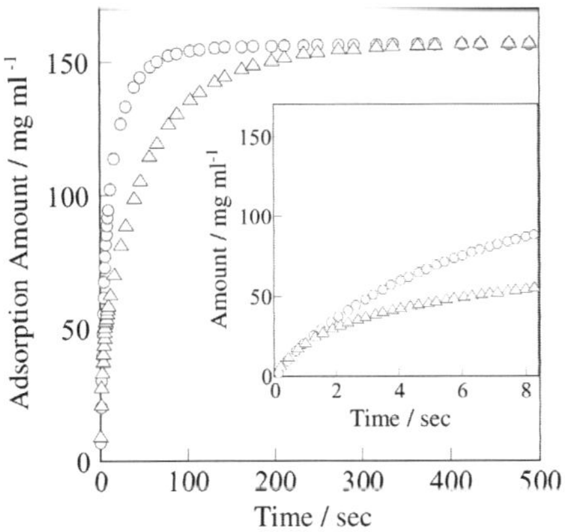

Figure 3. Adsorption rates of water molecules on ACFs of w = 0.7 (•) and 1.1 nm (Δ). Inset shows the adsorption rate in the shorter time-period.

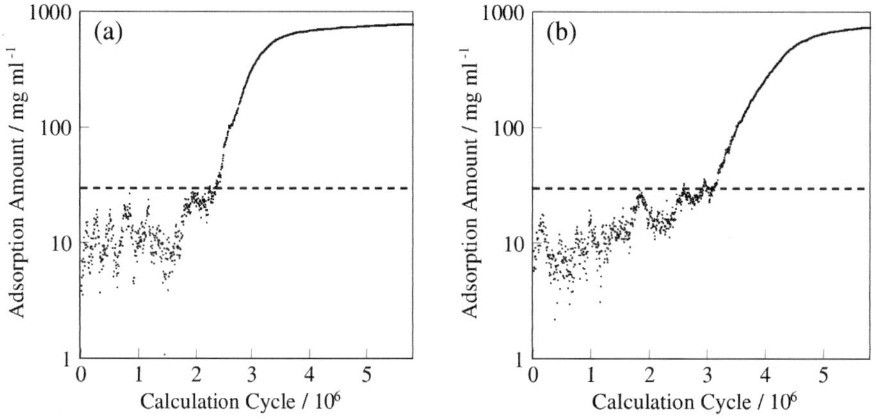

Figure 4. Simulated adsorption amount changes as function of calculation cycle for w = 0.7 (a) and 1.1 nm (b). The broken line indicates a critical adsorption amount.

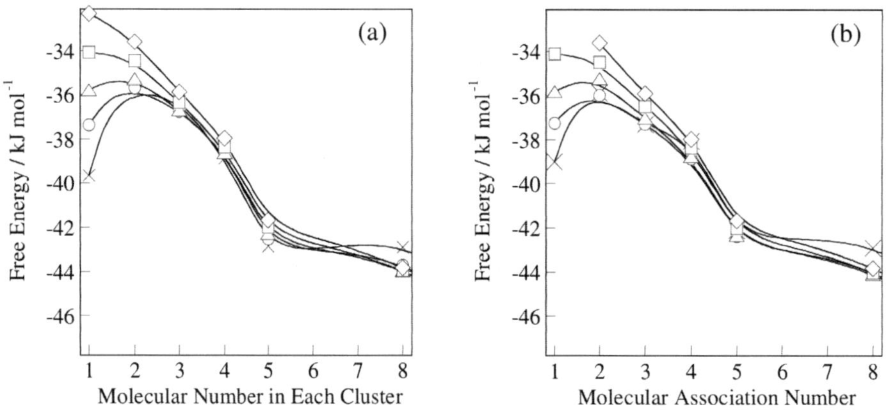

Figure 5. Free energy against molecular association number in each cluster as a function of simulated adsorbed density of 5 (×), 15 (•), 30 (△), 60 (□), and 120 mg ml^{-1} (◇). w = 0.7 (a) and 1.1 nm (b).

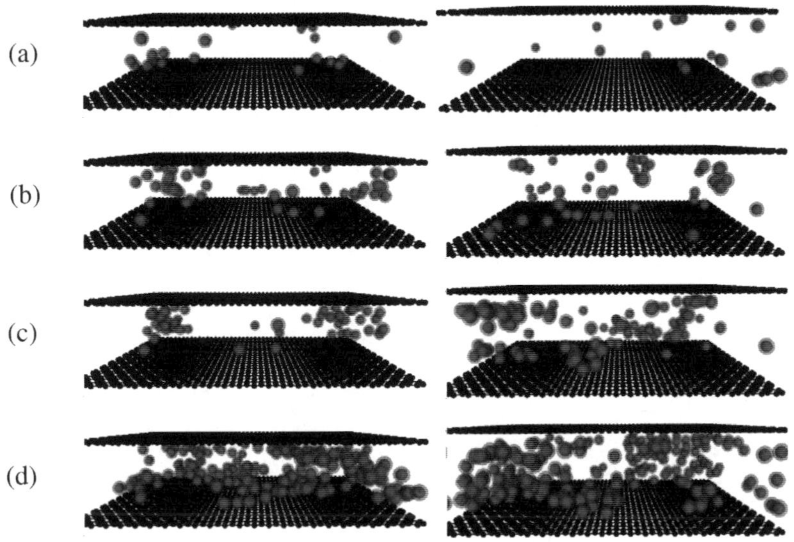

Figure 6. Snapshots of water molecules adsorbed in carbon nanopores of w = 0.7 (left) and 1.1 nm (right) at simulated amount of 15 (a), 30 (b), 60 (c), and 120 mg ml^{-1} (d). A water molecule is expressed by a large blue sphere and two small red spheres. A carbon atom on a graphitic sheet is colored in black.

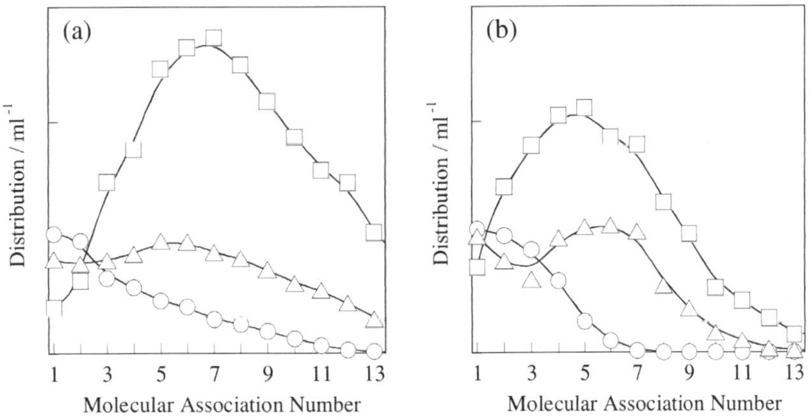

Figure 7. Association number distributions of water molecular clusters in pores of w = 0.7 (a) and 1.1 nm (b) at simulated amount of 30 (•), 60 (△), and 120 mg ml^{-1} (□).

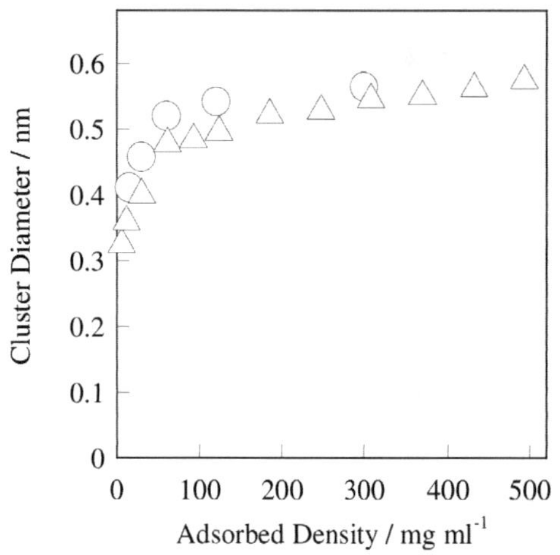

Figure 8. Adsorbed density dependence of mean cluster diameter.

•: w = 0.7 nm and Δ: w = 1.1 nm.

Computer simulation study of water/hydrocarbon interfaces: effects of hydrocarbon branching on interfacial properties

Janamejaya Chowdhary and Branka M Ladanyi[1]
Department of Chemistry
Colorado State University
Fort Collins, CO 80523-1872, USA.

E-mail: Branka.Ladanyi@colostate.edu, janamej@lamar.colostate.edu

Abstract. We review here the results of our molecular dynamics (MD) simulation study of water/hydrocarbon liquid/liquid interfaces. In order to examine the effects of chain length and branching on interfacial properties, we considered five different alkanes (n-pentane, 2-methyl pentane, 2,2,4-trimethyl pentane, 2-methyl heptane, and n-octane) as the hydrocarbon phase. We used a recently-proposed procedure to identify molecular surface sites and constructed intrinsic density profiles, in addition to the more familiar laboratory-frame profiles, in order to determine the effects of interface fluctuations on the structure and dynamics of the two phases. We found that interfacial properties of the aqueous phase are relatively insensitive to the molecular structure of the hydrocarbon, even though both branching and chain length have significant influence on the hydrocarbon interfacial properties. We found that translational and rotational mobilities of molecules of both phases are affected by the presence of the interface and that rotational relaxation of water molecules is significantly more anisotropic in the interfacial region than in the bulk.

1. Introduction

Liquid/liquid interfaces are ubiquitous in nature and are technologically important for processes such as surfactant self-assembly, oil recovery, and phase-transfer catalysis. Oil/water interfaces are among the most familiar examples of this type and represent a prototypical manifestation of hydrophobicity on an extended length scale. Computer simulation provides a great deal of insight into the molecular origin of the properties of liquid interfaces.[1] Several alkane/water liquid/liquid interfaces have been studied by molecular simulation, usually focusing on the interface of water with a particular alkane species[1,2] [2,3]. Our work was undertaken in part in order to examine how the properties of the aqueous phase depend on the molecular structure of the adjacent hydrocarbon phase [4,5]. We were also interested [6]in investigating how the

[1] To whom any correspondence should be addressed.

molecular relaxation processes in the interfacial region differ from the bulk phases and, in the case of water, how these differences are related to the interfacial hydrogen bond connectivity and lifetimes [5]. Another focus of our study was to use water/alkane systems to gain insight into the properties of the capillary waves arising at interfaces of immiscible liquids [7].

In this article, some of our main findings about the properties of water/alkane interfaces are reviewed, with the emphasis on the properties of the aqueous phase. Details of our work on water/alkane interfaces have either been published [4,5] or submitted for publication [8] elsewhere. We start by describing the model systems in Sec. 2. Structural features of these systems are presented in Sec. 3 and the results on the mobility of the molecules in the interfacial region in Sec. 4. The article is concluded in Sec. 5.

2. Model and simulation details

In this work we used effective potentials of the Lennard-Jones + Coulomb type to represent interaction sites of the molecules. In the case of water we used the SPC/E model [9] and for the alkanes the united-atom OPLS potential [10] in which the methyl and methylene groups are represented as single carbon-centered sites. Molecular dynamics (MD) simulations of five water/alkane systems under ambient conditions were carried out as described in Refs. [4,5]. Information on the specific systems studied is summarized in Table I. One of the goals of our work was to determine how the properties of the interfaces are affected by the extent of hydrocarbon branching and the chain length. For this purpose, we have classified the alkanes studied into two series: The molecules in Series 1 (n-pentane, 2-methyl pentane (2MP), and 2,2,4-trimethylpentane (TMP)) have the same chain length and different amounts of branching. Molecules in Series 2 (n-octane, 2-methyl heptane (2MH) and TMP) have the same mass and different amounts of branching. Our MD data were analyzed to determine the structural features of these alkane/water interfaces and the mobilities of the molecules in the interfacial region. In the case of water, these studies included the properties of the hydrogen bond network in the interfacial region.[4,8]

Table 1. Simulation setup[a]

Alkane	No. of mols.	Series	$\langle \rho \rangle$ (g/cm^3)
n-pentane (P)	91	1	0.625
2-methyl pentane (2MP)	79	1	0.650
2,2,4-trimethyl pentane(TMP)	64	1, 2	0.692
2-methyl heptane (2MH)	64	2	0.698
n-octane (O)	64	2	0.702

[a] The number of water molecules is fixed at 586. The temperature and pressure are maintained at 1 atm and 298 K.

3. Interfacial Structure

A simple way to characterize the interfacial structure is in terms density profiles of the molecules of the two phases in the 'laboratory' frame, i.e., the frame defined by the simulation cell. Figure 1 depicts these profiles, based on the locations of the water oxygen and alkane carbon sites, in the case of the water/TMP system. As can be see from this figure, the laboratory density profiles, especially the one of water, exhibit very little structure. This is a consequence in part of the fact that the liquid surfaces are not perfectly flat, but deviate from planarity due to the presence of capillary waves [11]. The extent of broadening of the density profiles due to capillary waves can be gauged by identifying the interfacial molecules, i.e., molecules for which the nearest neighbour belongs to the other phase. The distribution, which is approximately Gaussian, along z of these molecules is shown as dashed lines in figure 1.

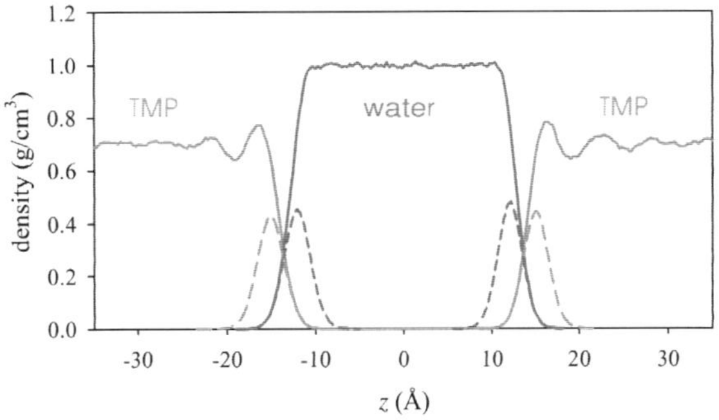

Figure 1. Density profiles for the water/TMP system. The density profile of molecules in the interfacial region is shown as dashed lines.

In order to determine the intrinsic structure from which these broadening effects have been removed, we follow the procedure developed by Pandit et al.[3] to locate the intrinsic interface. This procedure identifies the surface site relative to the location of the molecules of the other phase and uses Voronoi tessellation to determine the distance of molecules from the intrinsic surface location.

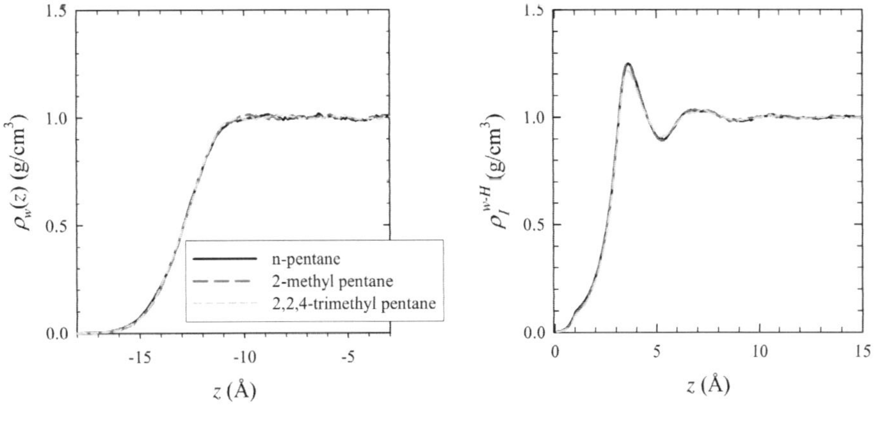

Figure 2. Water density profiles for water/alkane Series 1 interfaces. Left: density profile in the laboratory frame; right: intrinsic profile relative to the hydrocarbon phase.

As figure 2 illustrates in the case of water in the water/Series 1 alkane systems, the intrinsic density profile of water relative to the hydrocarbon surface exhibits much more pronounced interfacial structure than the corresponding laboratory profile. Figure 2 also illustrates the weak dependence of the water interfacial structure on the extent of branching in the hydrocarbon phase. The intrinsic profile $\rho_I^{w-H}(z)$ resembles water structure next to a solid hydrophobic wall [12].

Figure 3 shows that the density profile of the hydrocarbon phase relative to water, $\rho_I^{H-w}(z)$, exhibits more pronounced interfacial structure that the laboratory profile $\rho_H(z)$. The profiles for different alkanes differ substantially from each other, as shown in Figure 3 for the Series 1 liquids. Increased branching at fixed chain length leads to less efficient packing, as evidenced by a broader first peak and a larger distance between first and second peaks of the corresponding $\rho_I^{H-w}(z)$. These effects are less clear in the case of $\rho_H(z)$, for which increased branching mainly leads to larger peak amplitudes.

As water molecules approach the interface, they cannot participate in as many hydrogen bonds as in the bulk liquid. Next to a flat hydrophobic surface, some of the O-H bonds are approximately normal to the interface, with the H pointing towards the other phase.[12] We have shown that this is also true for the intrinsic O-H orientations at the liquid water/alkane interfaces.[4] These results are not shown here, but we illustrate in figure 4 how the average number of hydrogen bonds varies with the distance from the interface.

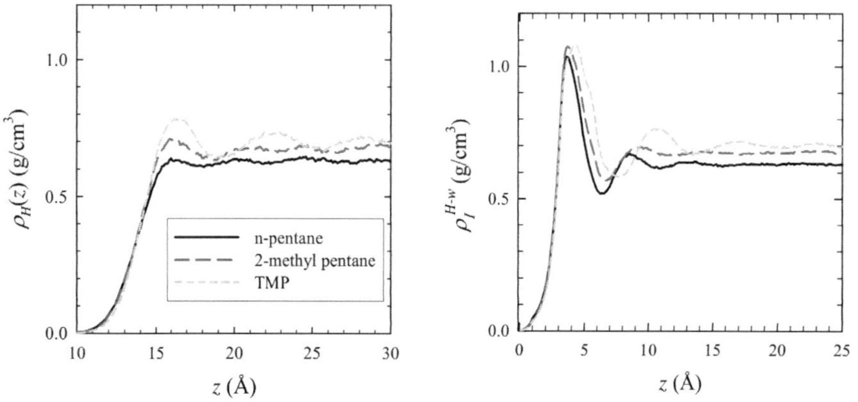

Figure 3. Alkane density profiles for water/alkane Series 1 interfaces. Left: density profile in the laboratory frame; right: intrinsic profile relative to the water phase.

As can be seen from figure 4, the extent of hydrogen bonding within the aqueous phase is essentially independent of the molecular structure within the hydrocarbon phase. There are, however, significant differences in the hydrogen bond profiles, $N^{HB}(z)$ and $N_I^{HB}(z)$ in the laboratory and intrinsic frames. In the intrinsic frame, the range of distances over which the average number of hydrogen bonds decreases from its bulk value is significantly shorter and reaches a larger value than in the laboratory frame. As in the case of density profiles, $N_I^{HB}(z)$ resembles more closely the behaviour of water next to a flat hydrophobic interface than does $N^{HB}(z)$.

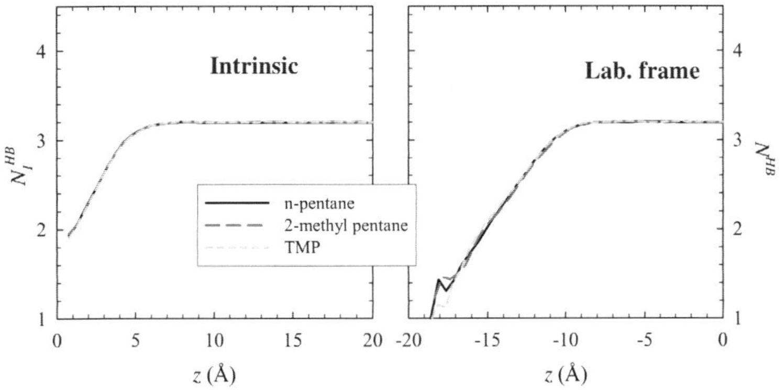

Figure 4. Average number of water-water hydrogen bonds as a function of z for water in the presence of Series 1 alkanes. Left panel: intrinsic hydrogen bond profile against distance from the alkane phase; right panel: laboratory-frame hydrogen bond profile.

4. Dynamics in the interfacial region

We have investigated the interfacial mobilities of the molecules of the two phases and their residence times in different interfacial layers. For this purpose, we have used both the intrinsic and laboratory-frame density profiles to determine the locations of molecules relative to the interface. The procedure used to identify the three regions (interfacial, intermediate and bulk-like) is illustrated for the laboratory-frame density profile of water. In the case of the intrinsic profile, the interfacial and intermediate regions corresponded approximately to the same ranges of z-values [5]

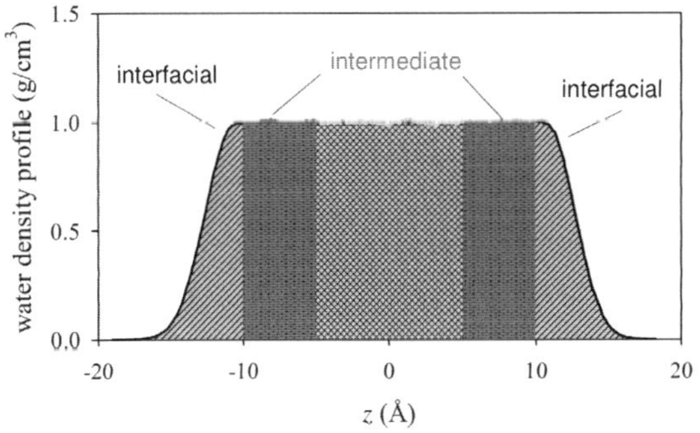

Figure 5. The water laboratory-frame density profile with the interfacial, intermediate, and core (bulk-like) regions displayed in different colors and patterns.

We refer the readers to our paper on interfacial dynamics for the results on the distributions of residence times and survival probabilities of the molecules in the different regions [5].

Mean squared displacements for molecules in the plane of, $\langle [\Delta r_{xy}(t)]^2 \rangle$, and perpendicular to, $\langle [\Delta r_z(t)]^2 \rangle$, the interface represent a simple measure of the translational mobility of molecules. In figure 6, these are shown for water in the water/TMP system in the laboratory frame. These results show that the translational mobility of water is enhanced in the interfacial region, leading to considerably faster in-plane diffusion. Given that the tendency of water to cross the interface towards the other phase is strongly suppressed, the longer-time slope of $\langle [\Delta r_z(t)]^2 \rangle$ is actually somewhat smaller at the interface than further away from it, despite the higher short-time mobility. Figure 6 also shows that diffusion in the intermediate region differs very little from the bulk. Thus we focus on the interfacial region in our further examination of translational mobility.

Journal of Physics: Conference Series **177** (2009) 012002 doi:10.1088/1742-6596/177/1/012002

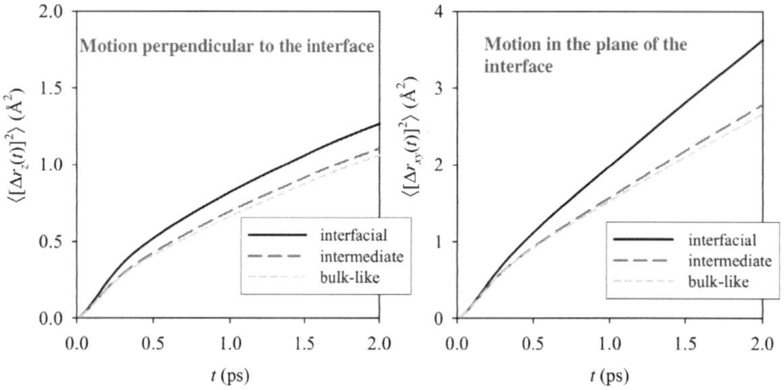

Figure 6. Mean squared displacements in different interfacial regions in the laboratory frame for water in the water/TMP system.

In figure 7 are shown the in-plane mean squared displacements for water in the interfacial regions of Series 1 and 2 water/alkane interfaces. The dependence of $\langle [\Delta r_{xy}(t)]^2 \rangle$ on the molecular structure of the alkane phase is weak, but shows definite trends with branching and chain length. Within Series 1, increased branching at fixed chain length leads to lower water mobility. Within Series 2, the mobility decreases with increasing chain length.

Also fairly modest is the difference in water translational mobilities when computed in the laboratory and intrinsic frames. This is illustrated in figure 8, where the results for $\langle [\Delta r_z(t)]^2 \rangle$ in water/(Series 1)alkane systems is displayed. We see that the mean squared displacements are very slightly higher in the intrinsic frame and that in both frames, the translational mobility normal to the interface decreases with increasing alkane branching, following the same trend as for in-plane translation.

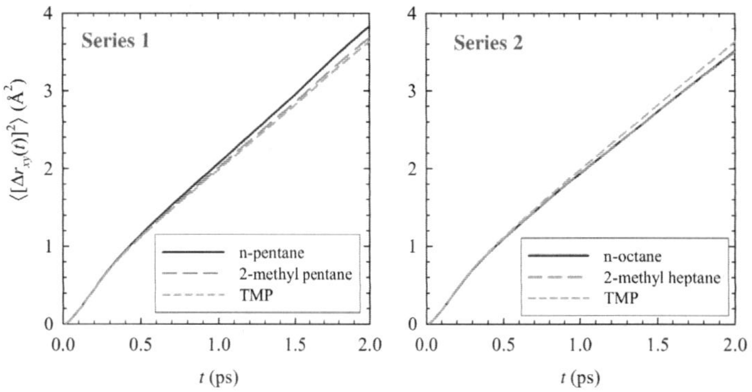

Figure 7. Mean squared displacements in the plane of the interface in the laboratory frame for water in the interfacial region in Series 1 (left) and Series 2 (right) water/alkane systems.

In addition to translational diffusion, orientational relaxation of molecules near the interface is expected to differ from the same process in the bulk liquid. We have investigated these differences by examining how they affect the orientational time correlations

$$C_l(t) = \langle P_l [\hat{\mathbf{u}}(0) \cdot \hat{\mathbf{u}}(t)] \rangle , \quad (1)$$

where P_l is a Legendre polynomial of order l and $\hat{\mathbf{u}}$ a unit vector embedded in a molecule. In the case of water, we chose unit vectors along the three principal axes of inertia.

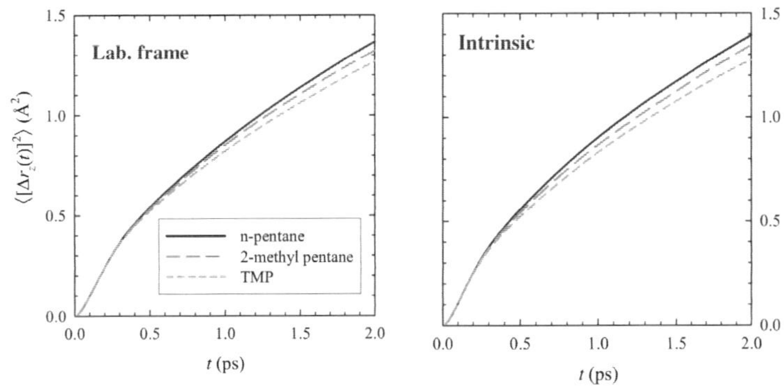

Figure 8. Water mean squared displacements perpendicular to the plane of the interface in the interfacial region for the Series 1 water/alkane systems. The left panel is for the laboratory and the right for the intrinsic frame.

The results for the relaxation of the time correlations $C_1(t)$ for the three vectors in the water/TMP system in the interfacial and bulk-like regions are shown in figure 9. As can be seen from the figure, reorientation of water is more anisotropic near the interface than in the bulk liquid, with a significant speed-up near the interface of the relaxation of the H-H and out-of-plane vectors, while the dipole relaxation becomes somewhat slower.

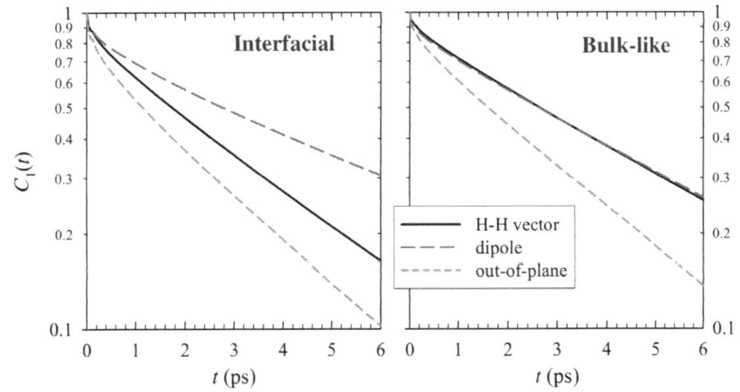

Figure 9. Orientational l=1 time correlations for the three principal axe of water in the water/TMP system in the laboratory frame. The left panel shows $C_1(t)$ in the interfacial and the right panel in the bulk-like region.

The results for the water/TMP system are representative of all the water/alkane systems we studied. A comparison of the results for the dipole vector relaxation in interfacial region of all the systems, displayed separately for Series 1 and 2, is shown in figure 10. These data indicate very weak dependence of $C_{1,\mu\mu}(t)$ on the molecular structure of the alkane phase. Within Series 1, a slight slowdown in the relaxation rate is seen as the extent of alkane branching increases.

We have also studied [5] interfacial orientational relaxation in the intrinsic frame and found that these results resemble closely the results for the laboratory frame. These results are reported in Ref. [5], but are not displayed here. The comparison of the results for the laboratory and intrinsic frames for water reorientation exhibits the level of agreement similar to that displayed in Figure 8 for the molecular mean squared displacements.

The most significant aspect of interfacial rotational mobility that our results display concerns the increased anisotropy in molecular reorientation in the interfacial region compared to the bulk liquid. We can gain further insight into this aspect of water interfacial mobility by examining the behavior of the time correlation of the hydrogen bond population in water/alkane interfacial systems. We have obtained the results for $C_{HB,k}(t)$, the time correlation for the fluctuations in the numbers of H-bonds in region k:

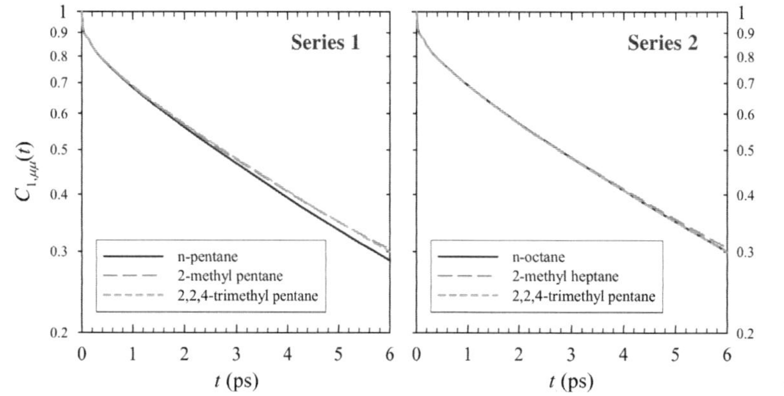

Figure 10. Orientational l=1 time correlations for the unit vector along the water dipole in the interfacial region of water/Series 1 alkanes (left) and water/Series 2 alkanes (right). in the laboratory frame.

$$C_{HB,k}(t) = \frac{\langle [h_{ij}(0) - \bar{h}_k][h_{ij}(t) - \bar{h}_k] \rangle}{\langle [h_{ij}(0) - \bar{h}_k]^2 \rangle}, \tag{2}$$

where $h_{ij}(t) = 1$ if there is a hydrogen bond between molecules i and j at time t and 0 otherwise.

$$\bar{h}_k = \left\langle \sum_{i \neq j \in k} h_{ij} / N_k^2 \right\rangle \tag{3}$$

is the average number of H-bonds and N_k is the number of molecules in region k.

Figure 11 displays side by side $C_{1,\mu\mu,k}(t)$ and $C_{HB,k}(t)$ in the three regions. The figure

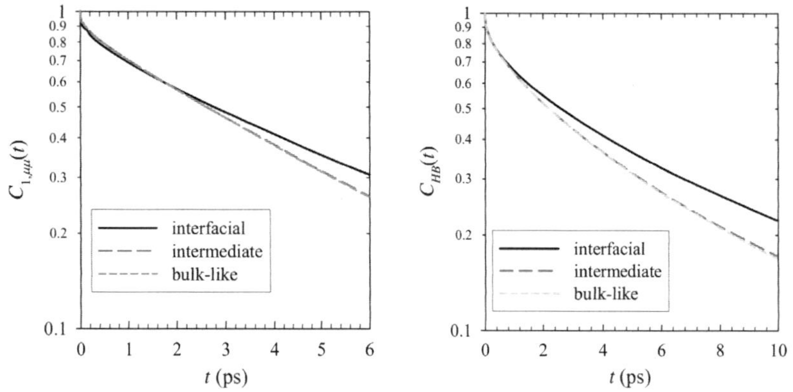

Figure 11. A comparison of the hydrogen bond (left) and dipole orientational (right) correlations in the different interfacial layers of the water/TMP system.

indicates that the two time correlations exhibit the same trends as a function of the distance from the interface. In particular, their behavior suggests that slower relaxation of the water dipoles in the interfacial region is related to the longer H-bond lifetimes in this region.

The enhanced anisotropy in water reorientation at the interface is also related to the preferred molecular orientations relative to the interface and how they affect the three principal axes. This information is displayed in figure 12 for the dipole and H-H vectors.

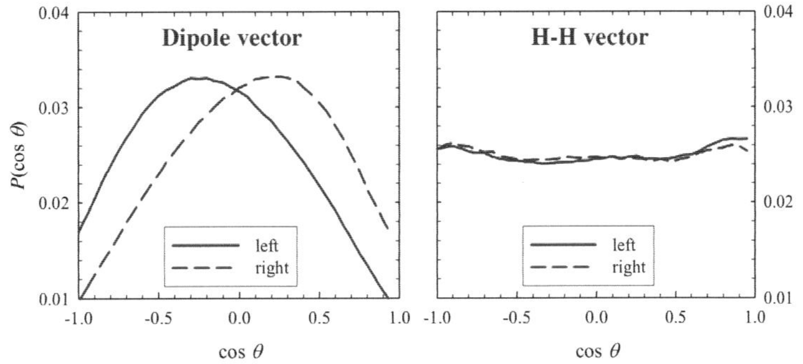

Figure 12. Probability distributions of $\cos\theta = \hat{\mathbf{u}}\cdot\hat{\mathbf{n}}$, where $\hat{\mathbf{n}}$ is the surface normal and $\hat{\mathbf{u}}$ a unit vector along a water molecule principal axis (left figure $\hat{\mathbf{u}}$ = along molecular dipole; right figure $\hat{\mathbf{u}}$ = along molecular H-H distance) in the interfacial region of water/TMP.

Specifically, in this figure are depicted the distributions of $\cos\theta = \hat{\mathbf{u}}\cdot\hat{\mathbf{n}}$, where $\hat{\mathbf{n}}$ is the surface normal along the positive z direction and $\hat{\mathbf{u}}$ a unit vector along a molecular principal axis. The distributions of for the dipole and H-H vectors are strikingly different from each other. The dipole clearly has a preferred orientation relative to the interface, with a peak at about 107°, predominantly in the plane of the interface, with a slight tilt towards the hydrocarbon phase, while the H-H vector has essentially no orientational preference. This indicates that the faster reorientation of the H-H vector at the interface than in the bulk is due mainly to the reduced water density in the interfacial region, while the slower reorientation is related to the difficulties of disrupting interfacial the H-bond network.

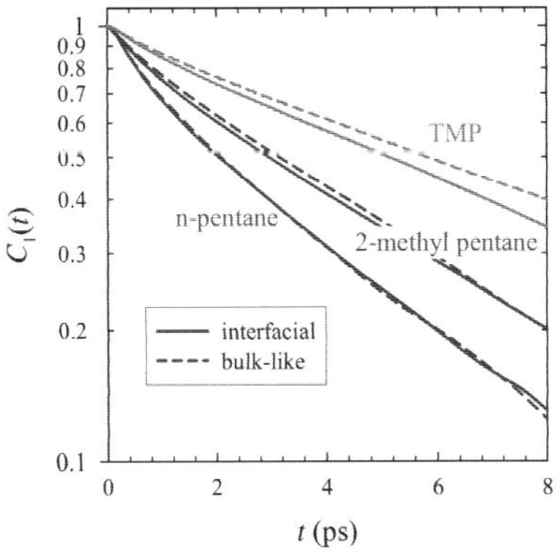

Figure 13. The orientational time correlations for the main-chain end-to-end vector of alkane molecules within Series 1 in the interfacial region and in the bulk.

While our focus has been primarily in characterizing the aqueous side of water/alkane interfaces, we have also studied the structural and dynamical properties of the alkane side of the interface. In figure 13 are shown the results for $C_1(t)$ of the vector connecting the main chain ends [4,5] for alkanes of Series 1. These results indicate that orientational relaxation rate is strongly dependent on the extent of branching within this series, whereas within Series 2, for which the results are not displayed here, slower relaxation rate is observed with increasing chain length. As can be seen from figure 13, the orientational relaxation rates of the three alkanes within Series 1 are only weakly dependent on the vicinity of the interface.

5. Summary and Conclusion

In this article, we have provided an overview of our MD simulation results for water/alkane liquid/liquid interfaces under ambient conditions. We have studied the properties of these interfaces for several different alkane phases in order to determine the effects of chain length and extent of branching. We have also studied how the interfacial structure and molecular mobility are affected by capillary waves. To this end, we determined the location of the intrinsic, nonplanar, interface using as a criterion the proximity of the molecules of the other phase [3]. We summarize here our main findings.

The diffuse nature of the density profiles in the laboratory-frame reflects the broadening due to capillary waves. The intrinsic density profiles reveal more pronounced structure. For the aqueous phase, this structure is essentially the same for all the alkane phases and resembles the structure near a planar hydrophobic wall [12]. By contrast, the intrinsic alkane density profiles are quite different from each other, reflecting the differences in molecular shapes and intermolecular correlations within this phase.

Water translational and rotational mobility is quite different near the interface compared to the bulk liquid. However, this difference does not persist deep into the aqueous phase. Molecules that are about 5 Å away from the interface exhibit essentially bulk-like dynamics.

Water interfacial dynamics is only weakly dependent on the molecular structure of the alkane phase. This conclusion is in general agreement with the experimental observations [6]. We have also found that interfacial dynamics are quite similar in the laboratory and intrinsic frames.

Water mobility near the interface reflects the fact that the presence of the alkane phase essentially blocks water motion in the direction towards this phase. Molecular mobility in other directions does not appear to be significantly hindered and is actually enhanced for some types of motion, such as translation in the plane of the interface.

Orientational relaxation of water molecules is significantly more anistropic near the interface than in the bulk. This is a consequence of the preferred molecular alignments relative to the interface and of the somewhat longer hydrogen bond lifetimes in the interfacial region.

In the case of the alkane phase, molecular mobility varies considerably with chain length and the extent of branching, but shows more modest effects due to molecular alignment in the interfacial region.

References

[1] Benjamin I. *Annu. Rev. Phys. Chem.* 1997; **48:** 407-51.
[2] Carpenter I L, Hehre W J. *J. Phys. Chem.* 1990; **94:** 531-6; van Buuren A R, Marrink S J, Berendsen H J C. *J. Phys. Chem.* 1993; **97:** 9206-12; Zhang Y H, Feller S E, Brooks B R, Pastor R W. *J. Chem. Phys.* 1995; **103:** 10252-66; Michael D, Benjamin I. *J. Phys. Chem.* 1995; **99:** 1530-6; Wang H B, Carlson E, Henderson D, Rowley R L. *Molecular Simulation* 2003; **29:** 777-85; Patel H A, Nauman E B, Garde S. *J. Chem. Phys.* 2003; **119:** 9199-206; Rivera J L, McCabe C, Cummings P T. *Phys. Rev. E* 2003; **67:** 011603; Jedlovszky P, Varga I, Gilayi T. *J. Chem. Phys.* 2004; **120:** 11839-51; Nicolas J P, de Souza N R. *J.*

Chem. Phys. 2004; **120:** 2464-9.
[3] Pandit S A, Bostick D, Berkowitz M L. *J. Chem. Phys.* 2003; **119:** 2199-205.
[4] Chowdhary J, Ladanyi B M. *J. Phys. Chem. B* 2006; **110:** 15442-53.
[5] Chowdhary J, Ladanyi B M. *J. Phys. Chem. B* 2008; **112:** 6259-73.
[6] Brown M G, Walker D S, Raymond E A, Richmond G L. *J. Phys. Chem. B* 2003; **107:** 237-44.
[7] Chowdhary J, Ladanyi B M. *Phys. Rev. E* 2008; **77:** 031609.
[8] Chowdhary J, Ladanyi B M. *J. Phys. Chem. B*, published on the web Nov. 20, 2008.
[9] Berendsen H J C, Grigera J R, Straatsma T P. *J. Phys. Chem.* 1987; **91:** 6269-71.
[10] Jorgensen W L, Madura J D, Swenson C J. *J. Am. Chem. Soc.* 1984; **106:** 6638-46.
[11] Weeks J D. *J. Chem. Phys.* 1977; **67:** 3106-21; Rowlinson J S, Widom B. Molecular Theory of Capillarity. Oxford, UK: Clarendon Press, 1982.
[12] Lee C Y, McCammon J A, Rossky P J. *J. Chem. Phys.* 1984; **80:** 4448-55.

Thermodynamics of supercooled water in solutions

D Corradini, P Gallo[1] and M Rovere

Dipartimento di Fisica, Università "Roma Tre"
Via della Vasca Navale 84, I-00146 Roma, Italy

E-mail: gallop@fis.uniroma3.it

Abstract. Molecular dynamics simulations are performed on bulk TIP4P water and on a low concentration aqueous sodium chloride solution. The thermodynamic properties of the systems are studied by analysing the isotherms and the isochores planes. The comparison of the thermodynamics planes of the two systems reveals that modifications beyond a trivial shift occur in the solution. Moreover the limit of mechanical stability and the temperatures of maximum density line can be calculated from the analysis of the thermodynamic planes. While the limit of mechanical stability results unaffected by the presence of the ions, the temperatures of maximum density line is both mildly shifted in temperature and modified in shape in the solution, with respect to bulk water. Signatures of the presence of liquid-liquid coexistence are found in the aqueous solution. The results are also compared to water in a hydrophobic environment and to water confined between hydrophobic plates.

1. Introduction and simulation details

Water presents several anomalous thermodynamic behaviours. Some of the most important are the presence of a line of temperatures of maximum density (TMD) in the thermodynamic plane and the divergences of thermodynamic response functions upon supercooling [1, 2].

The existence of a second critical point in the deep supercooled region of water, as the end point of the coexistence line between a low density liquid (LDL) and a high density liquid (HDL) has been hypothesized and its presence has been also observed in several molecular dynamics (MD) simulations [3–8]. In this framework, water thermodynamic anomalies are a consequence of long-range fluctuations induced by the second critical point. Although the deep supercooled region has been so far not directly accessible to experiments, signatures of the presence of a second critical point have been found also experimentally [9].

The properties of water as a solvent are also quite extraordinary [10]. In particular, aqueous solutions of electrolytes are interesting for their role in most biological and geophysical systems. Their thermodynamic properties upon supercooling are not well known so far and an improved understanding would be significant in several fields like for example in cryopreservation of biological tissues [11, 12].

Several important thermodynamics properties such as the melting point, the boiling point and viscosity are known to be affected by the presence of ions. It is therefore interesting to study whether and to what extent the presence of ions affects thermodynamic properties of water upon supercooling. Archer and Carter [13] have shown experimentally that for low NaCl content, up to concentrations

[1] To whom any correspondence should be addressed.

between 1 and 2 mol/kg at ambient pressure, the TMD and the anomalous behaviour of the isobaric specific heat are still present, while they disappear upon increasing salt concentration.

In this paper we present a MD study of TIP4P bulk water and of a sodium chloride solution in TIP4P water, denoted in the following NaCl(aq), with concentration c = 0.67 mol/kg [14]. Both systems are studied upon supercooling. We focus on the isotherms and the isochores planes of the two systems. The analysis of such planes allows the determination of the limit of mechanical stability (LMS) of the liquid with respect to the gas and of the TMD lines.

In our systems the particles interact via a potential given by the sum of a long-range coulombic potential and a short-range Lennard-Jones (LJ) potential. Therefore the interaction potential can be written as

$$U_{ij}(r) = \frac{q_i q_j}{r} + 4\varepsilon_{ij}\left[\left(\frac{\sigma_{ij}}{r}\right)^{12} - \left(\frac{\sigma_{ij}}{r}\right)^{6}\right] \tag{1}$$

where q are the electric charges, water-water LJ parameters are those of TIP4P model [15] and ion-ion and ion-water LJ parameters are taken from a reparametrization for LJ potential [16] of Pettitt and Rossky parameters [17] for the Huggins-Mayer potential. The long-range electrostatic interactions are treated by the Ewald summation method and the cutoff radius is set at 9.0 Å.

The MD simulations are performed in the *NVT* ensemble, employing Berendsen thermostat [18]. The integration timestep used is 1 fs. The simulation box contains 256 water molecules in the case of bulk water, while for the NaCl(aq) it contains 250 water molecules, 3 Na^+ ions and 3 Cl^- ions. We simulated densities in the range 0.80 g/cm^3 $\leq \rho \leq$ 1.05 g/cm^3 for bulk water and in the range 0.80 g/cm^3 $\leq \rho \leq$ 1.1 g/cm^3 for the NaCl(aq). For both systems the range of temperature spanned is 210 K $\leq$ T $\leq$ 500 K. The simulation procedure requires equilibration and production runs and the longest equilibration runs lasted up to 10 ns. We employed the DL_POLY simulation package [19]. Some of the state points for bulk water are taken from a previous work of our group [20].

2. Thermodynamic Results

The thermodynamic properties are studied comparing the isotherms and the isochores planes of bulk water and of NaCl(aq). In particular the TMD and the LMS lines of the two systems are also determined and compared. The LMS can be calculated from the divergence of the isothermal compressibility. The locus of the points for which

$$\left(\frac{\partial P}{\partial \rho}\right)_T = 0 \tag{2}$$

is the LMS line. Thus the line joining the minima of the isotherms defines the LMS line. In figure 1 and figure 2 the isotherms of bulk water and NaCl(aq) are shown respectively. Bulk isotherms in the P–ρ plane, figure 1, are reported for temperatures ranging from T = 210 K to T = 300 K. The T = 300 K isotherm is the first one showing a minimum, in correspondence of the density ρ = 0.83 g/cm^3. This minimum remains in correspondence to ρ = 0.83 g/cm^3 up to the T = 260 K isotherms and then shifts to ρ = 0.85 g/cm^3 for the lower temperatures isotherms. Also for the NaCl(aq) isotherms, figure 2, the first isotherm displaying a minimum is the T = 300 K, now in correspondence to ρ = 0.85 g/cm^3. For all lower temperatures isotherms the minimum is in correspondence to ρ = 0.87 g/cm^3.

Comparing the two isotherms planes we notice that the NaCl(aq) isotherms minima are shifted to higher densities, with respect to bulk water, and furthermore the envelope of the isotherm lines is globally shifted to lower pressures for the solution. For both systems we can observe that the T = 220 K and the T = 210 K isotherms present inflections crossing higher temperatures isotherms. This

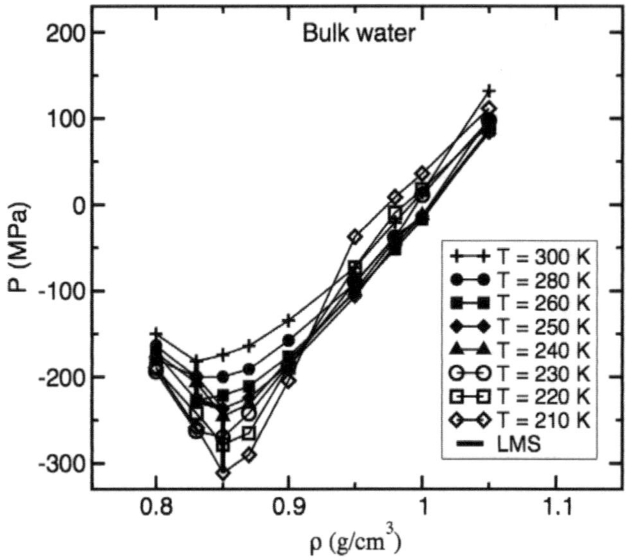

Figure 1. Isotherms of bulk water in the P–ρ plane and LMS line (thick line).

behaviour has been previously observed in bulk water [3, 21] and it has been interpreted as a signature of the approach of the system to the coexistence between LDL and HDL phases and eventually of the presence of a liquid-liquid (LL) critical point. Therefore it is possible to infer a LL phase coexistence and the existence of a second critical point also in our aqueous solution. As the inflections of NaCl(aq) isotherms are shifted to higher densities with respect to the bulk, we can hypothesize that also the critical point would be shifted to higher densities.

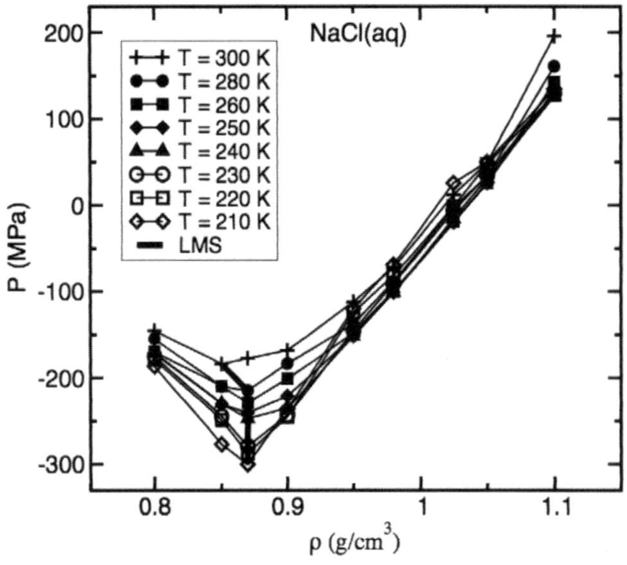

Figure 2. Isotherms of NaCl(aq) solution in the P–ρ plane and LMS line (thick line).

Journal of Physics: Conference Series **177** (2009) 012003 doi:10.1088/1742-6596/177/1/012003

The isochores of bulk water and NaCl(aq) are shown in figure 3 and figure 4 respectively. For both systems the isochores lying below the limit of mechanical stability are not reported. The TMD line can be extracted from the analysis of the isochores plane. The TMD line is the locus of the points where the coefficient of thermal expansion

$$\alpha_P = \frac{1}{V}\left(\frac{\partial V}{\partial T}\right)_P \tag{3}$$

goes to zero. As the the thermal pressure coefficient is given by

$$\gamma_V = \left(\frac{\partial P}{\partial T}\right)_V = \frac{\alpha_P}{K_T} \tag{4}$$

where K_T is the isothermal compressibility, the TMD line is then defined by the line that joins the minima of the isochores. Polynomial fits of the isochores were performed in order to determine more accurately the minima of the isochores (TMD points).

For bulk water, figure 3, only the four highest densities isochores exhibit a minimum, in the range of temperature explored. The shape of the TMD line for bulk water is in agreement with previous works [6, 21]. In the P–T plane of bulk water the LMS line is also reported and it almost coincides with the ρ = 0.85 g/cm^3 isochore. The LMS extends to the region of negative pressures and is nonre-entrant down to T = 210 K. The behaviour of the LMS line is analogous to what is found in the literature for several water models [4, 6, 21–26].

In figure 4 the isochores of the NaCl(aq) are shown along with the TMD and LMS lines. The isochores of the NaCl(aq) show minima down to the ρ = 0.90 g/cm^3 isochore. The presence of a TMD in a sodium chloride solution with the concentration studied is in agreement with the experimental work by Archer and Carter [13]. The LMS line for NaCl(aq) practically overlaps with the ρ = 0.87 g/cm^3 isochore.

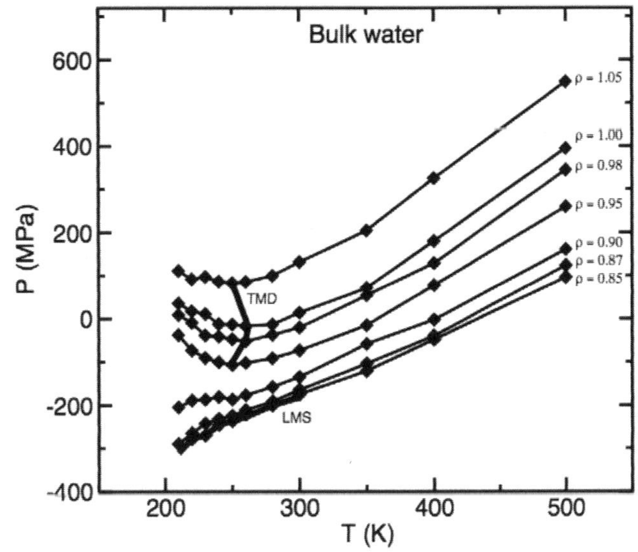

Figure 3. Isochores of bulk water in the P–T plane, TMD line LMS lines.

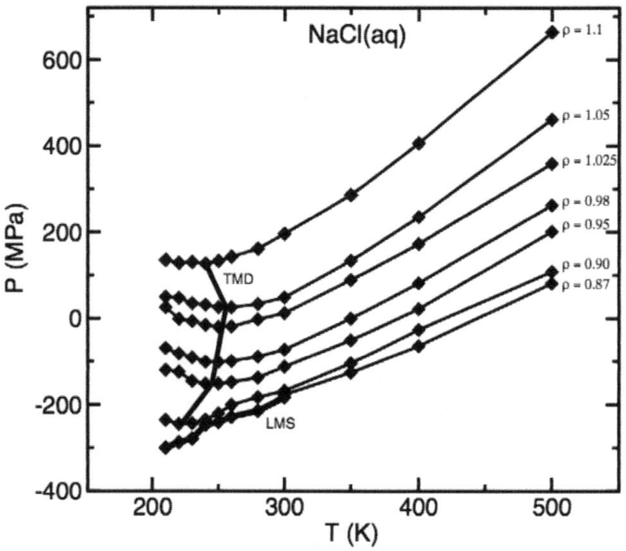

Figure 4. Isochores of NaCl(aq) solution in the P–T plane, TMD line LMS lines.

The comparison of the isochores planes of the two systems shows that the isochores of the aqueous solution lie at lower pressures with respect to bulk water and that the concavity of the curves close to the minima is markedly deeper in bulk water than in the solution. In figure 5 we report in the P–T plane, the LMS and the TMD lines of bulk water and of NaCl(aq). In the same picture we also show the LMS and TMD lines of TIP4P water confined in a hydrophobic environment, an off-lattice matrix of soft spheres [20]. Always in figure 5 the TMD lines of bulk TIP5P water [23] and of TIP5P water confined between hydrophobic plates [27] are reported.

Let us focus first on the effect of the ions on bulk water. We can notice that the presence of ions induces on the TMD line a mild shift in temperature and a marked modification in shape. Indeed, while the TMD lines of bulk water and of NaCl(aq) span the same range of densities, the range of pressures spanned by the NaCl(aq) TMD is much wider. The modification of the shape of the TMD together with shape modifications in the isotherms and isochores of the NaCl(aq), with respect to the bulk, indicate that the effect of the ions is not limited to give a trivial pressure shift in the phase diagram.

On the other hand, the presence of ions does not seem to affect the position and the shape of the LMS line that remains unaltered in the solution. We can argue that a possible explanation is connected to the fact that the position of the LMS line is mostly determined by the LDL phase. The ions, being hydrophilic, increase the local density of water, so that water around the ions is mostly HDL-like. The concentration of ions is probably too low to affect LDL properties and thus the LMS line does not change in the solution.

As we can see in figure 5, this is not the case for water confined in a hydrophobic environment [20] for which the LMS line is shifted toward higher pressures and reaches also higher temperatures, arriving at T = 350 K. This system resembles a solution of hydrophobic nonpolar solutes in water. Thus it can be argued that, as water density is depleted around hydrophobic solutes, the situation here is different with respect to the case of ionic (hydrophilic) solutes.

The TMD line of TIP4P water in the hydrophobic mixture is very similar to the one of the NaCl(aq) in shape, nonetheless it is shifted by about 220 MPa in pressure and about 20 K in

Journal of Physics: Conference Series **177** (2009) 012003 doi:10.1088/1742-6596/177/1/012003

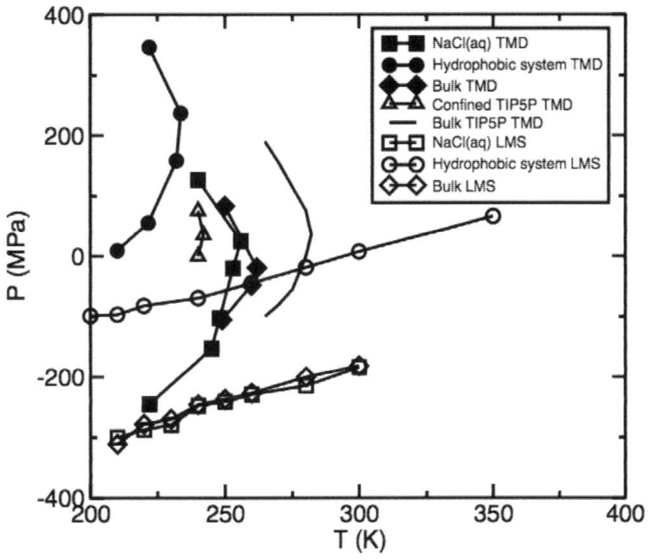

Figure 5. TMD and LMS lines in the P–T plane for bulk water, NaCl(aq), the hydrophobic system studied in [20] and TMD lines of TIP5P bulk water [23] and TIP5P water confined between hydrophobic plates studied in [27].

temperature. Therefore the presence of either hydrophilic or hydrophobic solutes in water seems to broaden the TMD line in a similar way, apart from the shift in the phase diagram due to the confinement. Moreover, the position of the TMD in confined systems depends on what of kind of confinement is applied. In fact, in TIP5P water confined by hydrophobic plates, Kumar et al. [27] found that the TMD line is shifted by about 40 K in temperature but they did not observe any pressure shift, with respect to bulk TIP5P water, see figure 5. For the system with hydrophobic soft spheres instead, the shift in temperature is similar but there are also a shift toward higher pressures by about 220 MPa and a broadening of the TMD line.

3. Conclusions

We performed MD simulations of bulk water and of a c = 0.67 mol/kg aqueous sodium chloride solution upon supercooling. In both cases the TIP4P model was used to model water molecules. Thermodynamic properties of the two systems were studied comparing the respective isotherms P–ρ planes and the isochores P–T planes. The presence of ions moves both the isotherms and the isochores toward lower pressures and slightly modifies the shape of the curves, especially in proximity of their minima. The analysis of these planes allowed also the determination of the TMD and the LMS lines. Our results were also compared to water in a hydrophobic environment [20], to TIP5P bulk water and to TIP5P water confined between hydrophobic plates [27].

The LMS line is substantially unaltered by the presence of ions, at the chosen concentration, with respect to bulk water, while it lies at higher pressures, reaching also higher temperatures for water in the hydrophobic environment. For all systems the LMS line extends to the region of negative pressures and it is nonre-entrant down to the lowest temperature examined. The different behaviour of the LMS lines in hydrophilic and hydrophobic systems could be connected by the local variation of the density of water induced by the presence of hydrophilic/hydrophobic solutes.

At variance with the LMS line, the TMD line is affected by the presence of the ions. In fact, it results mildly shifted in temperature and markedly modified in shape, with respect to the bulk, while it is both significantly shifted in temperature and affected in shape in the hydrophobic mixture. The TMD line is also deeply affected by the kind of (hydrophobic) confinement. While for water confined between hydrophobic plates only a shift in temperature was found, in water containing hydrophobic soft spheres a broadening of the TMD shape and a significant shift toward higher pressures are also present.

Signatures of the presence of LL coexistence were also found for the NaCl(aq) in the inflections of the low temperatures isotherms at high densities.

References

[1] Debenedetti P G 2003 *J. Phys.: Condens. Matt.* **15** R1670 and references therein
[2] Angell C A 1983 *Annu. Rev. Phys. Chem.* **34** 593
[3] Poole P H, Sciortino F, Essmann U and Stanley H E 1992 *Nature* **360** 324
[4] Tanaka H 1996 *J. Chem. Phys.* **105** 5099
[5] Brovchenko I, Geiger A and Oleinikova A 2005 *J. Chem. Phys.* **123** 044515
[6] Poole P H, Saika-Voivod I and Sciortino F 2005 *J. Phys.: Condens. Matter* **17** L431
[7] Jedlovszky P and Vallauri R 2005 *J. Chem. Phys.* **122** 081101
[8] Paschek D 2005 *Phys. Rev. Lett.* **94** 217802
[9] Mishima O and Stanley H E 1998 *Nature* **392** 164
ibid. 1998 **396** 329
[10] Angell C A 2002 *Chem. Rev.* **102** 2627
[11] Franks F 2000 *Water, a matrix for life* Royal Society of Chemistry
[12] Urrutia G et al. 2007 *Journal of Food Engineering* **83** 293
[13] Archer D G and Carter R W 2000 *J. Phys. Chem.* B **104** 8563
[14] Corradini D, Gallo P and Rovere M 2008 *J. Chem. Phys.* **128** 244508
[15] Jorgensen W L , Chandrasekhar J , Madura J D, Impey R W and Klein M L 1983 *J. Chem. Phys.* **79** 926
[16] Koneshan S and Rasaiah J C 2000 *J. Chem. Phys.* **113** 8125
[17] Pettitt B M, Rossky P J 1984 *J. Chem. Phys.* **84** 5836
[18] Berendsen H J C, Postma J P M, van Gunsteren W F, DiNola A and Haak J R 1984 *J. Chem. Phys.* **81** 3684
[19] Smith W, Forester T R and Todorov I T 2006 *The DL_POLY 2.0 User Manual* Daresbury Laboratory, Daresbury, UK
[20] Gallo P and Rovere M 2007 *Phys. Rev.* E **76**, 061202
[21] Poole P H, Sciortino F, Essmann U and Stanley H E 1993 *Phys. Rev.* E **48** 3799
[22] Sciortino F, Poole P H, Essmann U and Stanley H E 1997 *Phys. Rev.* E **55**, 727
[23] Yamada M, Mossa S, Stanley H E and Sciortino F 2002 *Phys. Rev. Lett.* **88** 195701
[24] Gallo P, Minozzi M and Rovere M 2007 *Phys. Rev.* E **75** 011201
[25] Harrington S, Poole P H, Sciortino F and Stanley H E 1997 *J. Chem. Phys.* **107** 7443
[26] Netz P A, Starr F W, Stanley H E and Barbosa M C 2001 *J. Chem. Phys.* **115** 344
[27] Kumar P, Buldyrev S V, Starr F W, Giovambattista N and Stanley H E 2005 *Phys. Rev.* E **72** 051503

Transferability of Polarizable Models for Ion–Water Electrostatic Interaction.

Marco Masia

Dipartimento di Chimica, Università degli Studi di Sassari, Sardinian Laboratory for Computational Materials Science SLACS (INFM-CNR) and INSTM, Via Vienna 2, 07100 Sassari, Italy.

E-mail: E-mail: marco.masia@uniss.it

Abstract. Studies of ion–water systems at condensed phase and at interfaces have pointed out that molecular and ionic polarization plays an important role for many phenomena ranging from hydrogen bond dynamics to water interfaces' structure. Classical and ab initio Molecular Dynamics simulations reveal that induced dipole moments at interfaces (e.g. air–water and water–protein) are usually high, hinting that polarizable models to be implemented in classical force fields should be very accurate in reproducing the electrostatic properties of the system. In this paper the electrostatic properties of three classical polarizable models for ion–water interaction are compared with ab initio results both at gas and condensed phase. For Li^+–water and Cl^-–water dimers the reproducibility of total dipole moments obtained with high level quantum chemical calculations is studied; for the same ions in liquid water, Car–Parrinello Molecular Dynamics simulations are used to compute the time evolution of ionic and molecular dipole moments, which are compared with the classical models. The PD2-H2O model developed by the author and coworkers [Masia et al. *J. Chem. Phys.* **2004**, 121, 7362] together with the gaussian intermolecular damping for ion–water interaction [Masia et al. *J. Chem. Phys.* **2005**, 123, 164505] showed to be the fittest in reproducing the ab initio results from gas to condensed phase, allowing for force field transferability.

1. Introduction

It is widely accepted that the implementation of polarizable force fields in classical Molecular Dynamics (MD) simulations is important for reproducing with high reliability the features of inhomogeneous systems [1, 2, 3, 4, 5, 6, 7]. Even though dipolar interactions decay faster than coulomb interactions they are known to be responsible of many macroscopic properties of interfaces [8, 9, 10, 11]. In particular, in water–air, water–protein and water–framework (in confined geometries) interfaces, the induced dipole moments vary over a wide range in different regions of the system and are responsible of many phenomena such as nanobubbles' coalescence [12, 13]; in the bulk, ionic and molecular polarization influences solvation shell dynamics [14, 15, 16, 17, 18, 19, 20]. For these reasons it is required that the polarizable force fields implemented in MD simulations are able to faithfully reproduce both the energetic and electrostatic properties without diverging for high electric fields. Most of the studies on polarizable models focus only on the energetic properties; as usual in force field development, parameters are optimized to fit the ab initio energies of a large sample of configurations. Only recently it has been observed that, although common force fields fit with high accuracy the energetic properties, they perform poorly in reproducing the ab initio induced dipole moments [21]. This perspective guided the author in pursuing a method to develop polarizable models which could faithfully reproduce the electrostatic properties of simple systems such as ion-molecule (either water or carbon tetrachloride) dimers [21, 22, 23, 24, 25]. In particular the focus of previous papers was on the validation of polarizable models usually implemented in MD simulations: classical calculations for simple systems at gas phase were compared to high level quantum chemical calculations. The rationale behind this approach is that the electrostatic properties (arising from the interactions among point charges and point dipoles) should be in agreement with the ab initio calculations; this is already the standard way for developing intermolecular no–polarizable force fields. Point charges are obtained by a fit of the ab initio electrostatic potential with the additional constraint of reproducing the permanent dipole moment of the molecules [26, 27, 28, 29].

In the case of polarizable models the approach proposed is to find a set of functions and parameters (*vide infra*) which fit the ab initio induced dipole moments for a certain number of sample configurations for dimers. Once a good polarizable model is found, the following step for the development of the complete force field should be, then, to parameterize short range interactions, what could be done with the common methods of force field parameterization (*e.g.* energy fitting of many sample configurations). Before that, the issue of the model portability should be addressed; in fact this feature is required for a successful implementation of a force field in the study of a wide range of systems from gas to condensed phase. Recently, many authors [30, 31, 23, 24, 25, 32, 33] have pointed out that one of the largest limitations of standard molecular-mechanics polarizable force fields lies in that the transferability of gas phase derived potentials to condensed phase is hindered by the absence of polarization exchange–coupling in classical models (since it is due to short range electron cloud repulsion, it is also known as *Pauli effect* [33]). The portability of water–ion polarizable models is the focus of this contribution. In previous works, the author and his coworkers showed that, among the most popular methods implemented in MD algorithms to include polarization, the induced Point Dipole method performs better than fluctuacting charges and Drude models. Moreover, in the case of ion-molecule electrostatic interaction, it has been shown that an intermolecular damping function should be added to account for short range electron repulsion which cause the effective polarizability to be lower. This effect is independent of the species polarizability and it is rather related to the Van der Waals radius; in fact it was found to be important both for ions with high (anions) and low polarizabilities (cations) [23, 24, 25]. The outline of this article is as follows: in section 2 we discuss the computational details and the methods used; results and conclusion are given respectively in sections 3 and 4.

2. Computational Details

2.1. Ab Initio Calculations

To explore the portability of the polarizable force field, the results of ab initio calculations for gas phase and condensed phase were compared with classical calculations. Reference ab initio data were obtained using different approaches depending on the size of the system.

In the case of ion–water dimers five different configurations were chosen; for each of them the ion–oxygen distance was varied within an interval of ~ 5 Å. The shortest distance was selected to be be equal to the distance where the potential energy was ≈ 25 kJ mol^{-1} ($\approx 10k_BT$ at standard temperature); this criterium is very strict since some points are very likely to be inaccessible in condensed phase [21, 22, 23]. This fact should considered when assessing the significance of the results at very short distances.

All quantum chemical calculations were performed with the commercial package Gaussian 03 [34]. The B3LYP/aug-cc-pvtz model chemistry was used for all atoms. The density functional used was choosen because it is known to perform well, with small errors for the computed dipole moments and polarizabilities [35]. In a previous work we estimated that for such systems B3LYP slightly overestimate CCSD calculations, the largest difference being $\sim 4\%$ [24]. The total dipole moment of the dimer has been defined taking the position of the ion as the origin of the reference system. In this way, in the limit of zero polarizability for the ion the total dipole moment of the dimer is equal to that of the molecule.

The pure quantum chemical approach for condensed phase system would not be feasible, being the size of the system too large; in fact, since the configuration space to sample would increase exponentially with the number of molecules, performing ab initio calculations for each point would require huge computational resources. To overcome this problem, the Car-Parrinello (CP) [36, 37] scheme for propagating the wavefunctions and the configurations has been used. This allows to sample a great number of configurations that could occur in typical MD simulations. The system under study was formed by a single ion surrounded by 64 water molecules at ambient conditions. The BLYP/PW (Plane Waves) model chemistry [38, 39] was used with Troullier–Martins pseudopotentials [40] with a 70 Ry cutoff. The time step was set to 4 a.u. and the fictitious mass for the orbital was chosen to be 400 a.m.u. Periodic boundary conditions were applied. Production runs of 30 ps followed equilibration runs of 10 ps. Every five time steps, the Wannier centers' [41, 42, 43] coordinates $\vec{r}_j(t)$ were computed (see figure 1); in this way, given the ionic coordinates $\vec{R}_i(t)$, the dipole moments of molecule (or ion) I were obtained as:

$$\vec{\mu}_I = \sum_{i=1}^{N_I} Q_i \vec{R}_i + \sum_{j=1}^{n_I} q_j \vec{r}_j, \tag{1}$$

Q_i and q_j being respectively the charge of the ith ion and of the jth Wannier center. Contrary to the case of ion–water dimers, the frame of reference chosen for charged species was the molecular (ionic) center of mass. The advantage of this approach as compared to the one used for dimers is that the contribution of each species to the ab initio dipole moment can be known. On the other hand the approach used for dimers gives high quality results (which are needed for a fine tuning of parameters) but are limited to the total dipole moment of the dimer.

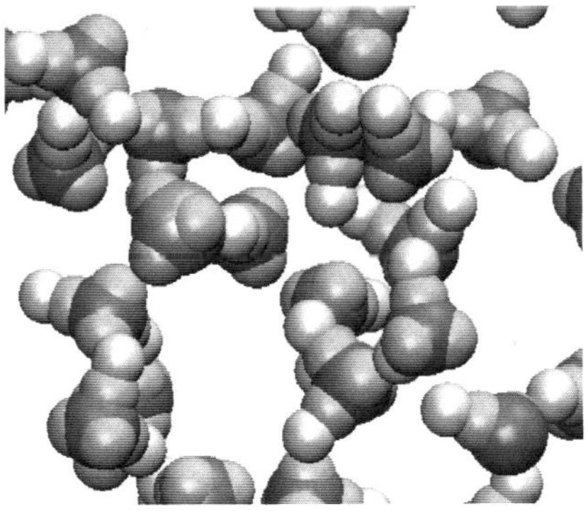

Figure 1. Detail of one configuration of chloride–water at condensed phase. Besides the atoms, Wannier Centers representing electron pairs are shown in green.

2.2. Classical Calculations

For all the configurations sampled with ab initio methods, the molecular (and ionic) dipole moments were calculated with the Point Dipoles (PD) method [44, 7]. Within this method, a value of polarizability α_j is assigned to site j; it is not required that such sites are localized on the atoms, neither that they bear physically meaningful values of α_j. The only requirement is that the polarizability tensor $\hat{\alpha}_I$ (which could be easily obtained from the site polarizabilities and their coordinates) of the whole molecule (ion) is comparable to the experimental one. The

Table 1. Polarizable models and optimized values for the Cl^-–water damping parameters.

WATER MODEL	SITE POLARIZABILITIES		DAMPING PARAMETERS			
	α_M (Å^3)	α_H (Å^3)	$a_{\mathrm{Cl-M}}$	$b_{\mathrm{Cl-M}}$	$a_{\mathrm{Cl-H}}$	$b_{\mathrm{Cl-H}}$
EXP-PDM-H2O	1.44400	0.00000	0.00	3.26	1.67	0.00
GAU-PDM-H2O	1.44400	0.00000	0.00	7.50	2.29	0.00
EXP-PD1-H2O	1.42048	0.00192	0.00	3.15	1.57	0.00
GAU-PD1-H2O	1.42048	0.00192	0.00	2.46	1.23	0.00
EXP-PD2-H2O	1.40990	0.00380	0.00	3.14	1.56	0.00
GAU-PD2-H2O	1.40990	0.00380	0.00	7.64	1.51	0.00

ions were assigned their polarizability at gas phase, while water molecules were described with three different models:

(i) PDM-H2O: one single polarizable site "M" is located at a distance $d = 0.215$ Å from the oxygen along the permanent dipole moment vector;

(ii) PD1-H2O: two site polarizabilities are assigned to the hydrogen atoms and one polarizable site "M" is located at a distance $d = 0.22$ Å from the oxygen along the permanent dipole moment vector;

(iii) PD2-H2O: two site polarizabilities are assigned to the hydrogen atoms and one polarizable site "M" is located at a distance $d = 0.0606$ Åfrom the oxygen along the permanent dipole moment vector.

Independently on the model, all sites bear the same charge; more details are given in table 1.

A detailed description of polarization methods was given in Refs. [21, 23]. Here the main features of the PD method are outlined: the total electric field ($\vec{E}$) acting on each polarizable site is obtained by the field $\vec{E}^q$ produced by the external partial charges and by the field $\vec{E}^\mu$ due to the intramolecular and external induced dipole moments:

$$\vec{E}_i^q = \sum_{j \neq i} T_j^q q_i \tag{2}$$

$$\vec{E}_i^\mu = \sum_{j \neq i} T_{ij}^\mu \cdot \vec{\mu}_i \tag{3}$$

$$\vec{E}_i = \vec{E}_i^q + \vec{E}_i^\mu, \tag{4}$$

where T_i^q and T_{ij}^μ denote respectively the charge and the dipole field tensor, which for a charge and a point dipole located at the origin can be written as:

$$T^q = \frac{\vec{r}}{r^3} \tag{5}$$

$$T^\mu = 3\frac{\vec{r}\vec{r}}{r^5} - \frac{I}{r^3}. \tag{6}$$

The induced point dipole on site i is obtained from the total field according to

$$\vec{\mu}_i = \alpha_i \vec{E}_i, \tag{7}$$

and is computed iteratively until a given threshold of convergence for the induced dipole is reached. It has been shown that for a realistic description of the short range interactions the electric field created by a fixed charge and/or that created by a point dipole should be damped by functions $f_q(r)$ and $f_\mu(r)$; eqs. 5 and 6 are substituted by the following:

$$T^q = f_q(r) \cdot \frac{\vec{r}}{r^3} \tag{8}$$

$$T^\mu = f_\mu(r) \cdot 3\frac{\vec{r}\vec{r}}{r^5} - f_q(r) \cdot \frac{I}{r^3}, \tag{9}$$

where, in the limit of *point* charges and/or *point* dipoles we recover the usual expressions, being $f_q(r) = f_\mu(r) = 1$. On the other hand, when charges and dipoles are thought to be spatially extended, the form of the damping depends on the charge distribution assumed; in ref. [23] the following functional forms were proposed:

(i) *linear damping*:

$$f_q(r) = 4\left(\frac{r}{a}\right)^3 - 3\left(\frac{r}{a}\right)^4 \tag{10}$$

$$f_\mu(r) = \left(\frac{r}{b}\right)^4, \tag{11}$$

(ii) *exponential damping*:

$$f_q(r) = 1 - e^{-(\frac{r}{a})^3} \tag{12}$$

$$f_\mu(r) = 1 - \left[1 + \left(\frac{r}{b}\right)^3\right] e^{-(\frac{r}{b})^3}, \tag{13}$$

(iii) *gaussian damping*:

$$f_q(r) = \text{erf}\left(\frac{r}{a}\right) - \frac{2}{\sqrt{\pi}}\left(\frac{r}{a}\right) e^{-(\frac{r}{a})^2} \tag{14}$$

$$f_\mu(r) = \text{erf}\left(\frac{r}{b}\right) - \frac{2}{\sqrt{\pi}}\left(\frac{r}{b}\right) e^{-(\frac{r}{b})^2}\left[1 + \frac{2}{3}\left(\frac{r}{b}\right)^2\right]. \tag{15}$$

In any of the above possibilites a and b play the role of the characteristic distance of maximum approach. Their value depends on the functional form chosen, and is dependent on each different pair.

Since the *linear* damping shows a pathological behaviour at $r = a$ or $r = b$ which might be problematic in Molecular Dynamics simulations, in this paper only the *exponential and gaussian* damping functions are explored in combination with the polarizable models introduced above. All the parameters used are shown in table 1 where, to the water models introduced we add the prefix EXP or GAU depending on the kind of damping function used. The damping parameters for lithium–water dimers are reported in ref. [23]. In the case of chloride–water dimers, the a and b parameters of eqs. 12, 13, 14 and 15 were optimized using the simplex algorithm [45]; the root mean square difference (RMSD) between ab initio and classical dipole moments for all configurations sampled was minimized. In this sense, those parameters represent the best choice which reproduce the ab initio results.

Journal of Physics: Conference Series **177** (2009) 012004 doi:10.1088/1742-6596/177/1/012004

3. Results and Discussion

3.1. Dimers

Lithium and chloride ions were chosen because of their electronic characteristics: Li^+ has a high charge density and a low polarizability ($\alpha = 0.02875$ Å^3) [46], while Cl^- has a lower charge density (being its Van der Waals radius bigger than Lithium) and a high polarizability ($\alpha_{Cl} = 5.4822$ Å^3) [47]. The two ions feature opposite qualities which are useful for testing the classical models against ab initio calculations.

The total dipole moment of the dimers is sketched in figures 2 and 3. The configurations showed (respectively for the lithium–water and chloride–water dimers), being energetically favoured both at gas and at condensed phase, are the most representative and they weight more than others for condensed phase simulations.

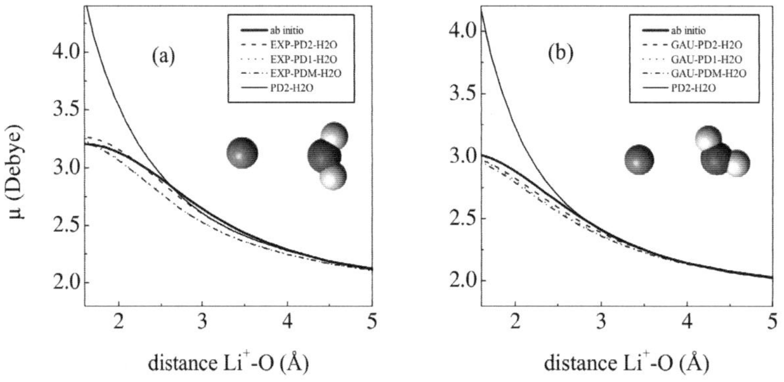

Figure 2. Total dipole moment of the lithium–water dimer as a function of the ion–oxygen distance for the *face* (left panel) and *trans* (right panel) configurations (see insets). (a) all models with exponential damping and without damping; (b) all models with gaussian damping and without damping.

It can be appreciated that the models without damping overestimate the ab initio curve at short distances. As explained above, this is due to the so called "Pauli effects" in the multi–body interactions [33]. The only way to reproduce the ab initio curve would be to vary the site polarizabilities both for water and for the ion. In MD simulations of halide–water at condensed phase, this is indeed the typical way to circumvent the "polarization catastrophe", *i.e.* the rise to infinity of the induced dipoles [50, 48, 49, 14]. Nevertheless, as shown in refs. [24, 25] the recourse to a lower polarizability causes the total dipole moment of the dimer to be underestimated at intermediate distances. Moreover, by changing site polarizabilities to hinder the model divergence, force field transferability would be affected, since it can not be assured that in different conditions it works out well.

The development of PDX-H2O models was done under the constraint that the gas phase molecular (ionic) polarizability should be reproduced; some model is even able to reproduce the water polarizability tensor. Accounting for short range electronic repulsion does not require to tune site polarizabilities; as can be seen in figures 2 and 3, in fact, the performance of all models is greatly increased by implementing the simple damping functions introduced in the previous section. In these cases the dipole moment at long and intermediate distances is still preserved,

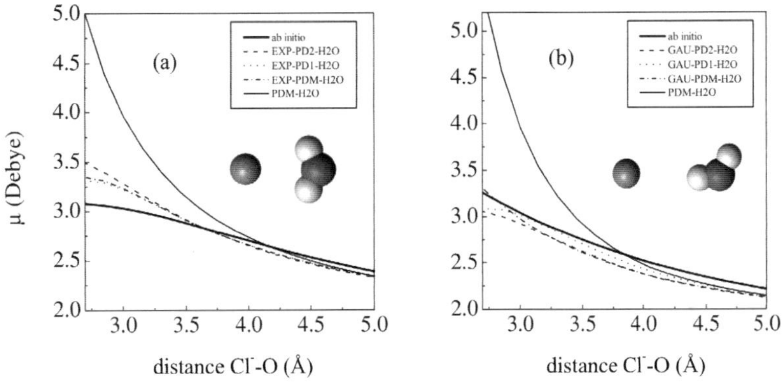

Figure 3. Total dipole moment of the chloride–water dimer as a function of the ion–oxygen distance for the *back* (left panel) and *cis* (right panel) configurations (see insets). (a) all models with exponential damping and without damping; (b) all models with gaussian damping and without damping.

and the short range part is reproduced with small errors. Some differences between all models can be appreciated at short distance.

For lithium–water (figure 2) it can be seen that the ab initio curve is fitted very well by all models in both configurations, and a better performance is obtained by the *XXX*-PD2-H2O models. Both the exponential and the gaussian dampings give comparable results.

On the other hand, for chloride–water, the performance of all models depends on the configuration considered. In the *back* configuration (panel *a*) the EXP-PD1-H2O and the GAU-PDM-H2O (not shown) models show the smallest displacement with respect to the other models; while in the *cis* configuration, the GAU-PD2-H2O and the EXP-PD1-H2O models are the best.

For both ions, though, it should be noticed that, overall, the differences among all models are not big and that they occur at (so short) distances where, as observed in section 2.1, it is very unlikely to access in MD simulations at standard conditions. Moreover those small differences cancel when other configurations (in addition to the most probable ones shown in figures 2 and 3) are considered. It can be concluded that all of the models proposed are good enough to reproduce the ab initio dipole moment of the dimers, for both ion–water dimers.

3.2. Condensed Phase

For condensed phase simulations it would not make sense to compare the ab initio total dipole moment of the system with the classical ones. In fact the contributions of many different molecular orientations would involve big cancelations and there would exist many parameter sets which faithfully reproduce the ab initio results. Moreover no insight could be obtained on the contributions of every single species in solution. By calculating the Wannier Centers, ionic and molecular dipole moments are easily computed (see section 2.1).

Figure 4 shows the time evolution of the dipole moment as computed from the GAU-PD2-H2O classical model (left panel) and from the Wannier Centers (right panel) for the ion and for one water molecule in the chloride–water system. It can be noticed that for both species the classical and ab initio results are almost superposable.

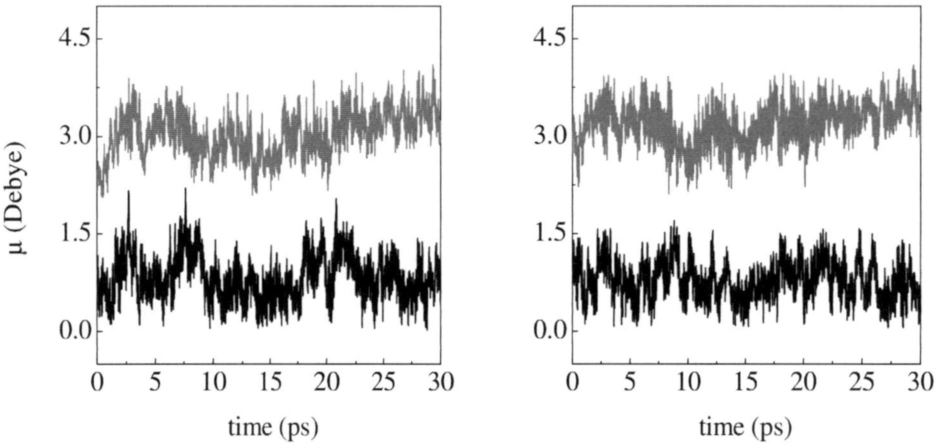

Figure 4. Time evolution for classical (left panel) and ab initio (right panel) dipole moments of th Cl^- ion and of one water molecule.

Further insight in the goodness of each model could be obtained from dipole–dipole correlation plots: in the case of first shell molecules in the lithium–water system, the comparison of PDX-H2O models with different damping functions is shown in figure 5. The PD2-H2O model perform much better than PD1-H2O and PDM-H2O (which, being almost superposable, can not be shown in the same figure); underestimation of ab initio results is common to other classical models recently published where the shift in the classical induced dipole moments rises up to 50% [51, 52].

The same conclusion is obtained when the dipole moment distribution functions $P(\mu)$ are considered (see figure 6). Either the EXP-PD2-H2O and the GAU-PD2-H2O models give good results both for first shell and bulk water molecules. The corresponding curves for Li^+ are peaked at 0.01 Debye both for ab initio and classical models. From the analysis of $P(\mu)$ it can be observed a small displacement of the PD2-H2O models with respect to ab initio calculations for first shell molecules which escaped from the study of the dipole–dipole correlation plot. Such small shift (around 0.3 Debye, less than 10%) could be due to the fact that water–water interactions are not damped. Li^+ solvation shell has in fact a tetrahedral geometry where water molecules occupy each vertex at an average distance of ≈ 2 Å and are rigidly oriented towards the ions [53]. The fact that first shell molecules' dipole moment is so highly influenced highlights that no–polarizable models would miss important electrostatic contributions and that accounting for polarizable interactions is needed also for ions with low (almost zero) polarizability. For what concerns bulk water molecules, it seems that they are not polarized by the ion, since their average dipole moment agrees with what was found in previous calculations [54] of pure liquid water.

A similar (but smaller) shift in the point dipole distribution functions of first shell molecules is found for chloride–water (see figure 7). Also in this case both exponential and gaussian damping functions are almost equally good in reproducing water induced dipole moments.

On the other hand, if we look at the induced dipole moment on the chloride ion, it can be observed that the exponential damping fails, since it gives a much broader distribution peaked

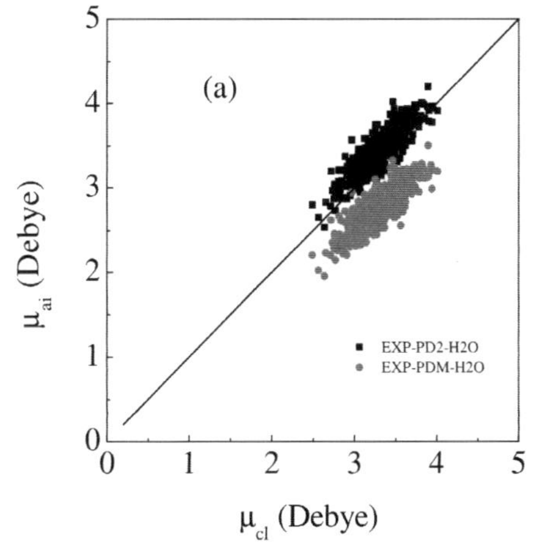

Figure 5. Correlation plot of ab initio (μ_{ai}) against classical (μ_{cl}) dipole moments for first shell molecules for Li^+ in water. Panel (a) and (b) show respectively the results for exponential and gaussian dampings with different water models.

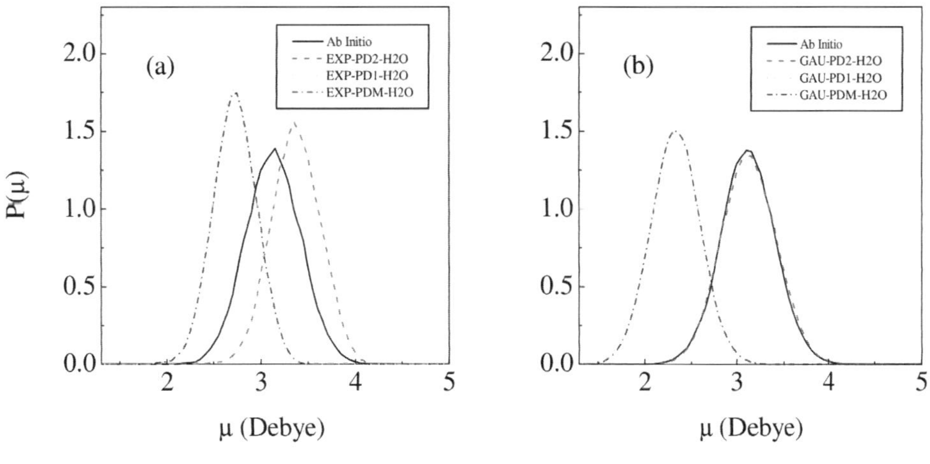

Figure 6. Dipole moment distribution for (a) first shell and (b) bulk shell water molecules from ab initio and classical calculations of Li^+ in water.

at higher values of μ; only by implementing the gaussian damping it is possible to reproduce with high precision the ab initio distriution. This is in contrast to what was found for the dimer,

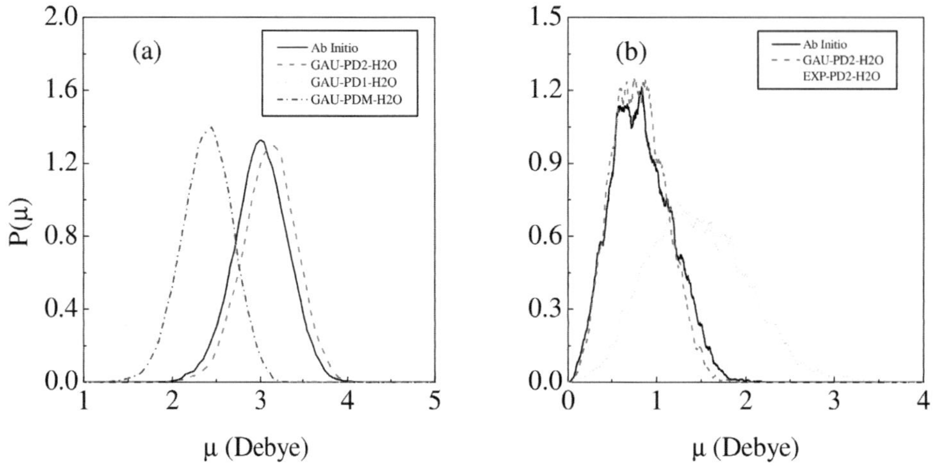

Figure 7. Dipole moment distribution for (a) first shell water molecules and (b) chloride anion from ab initio and classical calculations of Cl^- in water.

where both kind of damping functions could be used to model the Cl^-–water system.

4. Concluding Remarks

The use of intermolecular damping functions for modeling ion–water polarizable interaction is of great importance for accounting for short range electron cloud repulsion. This is evident already for the dimer at gas phase; in this case the PD2-H2O water model with the two types of damping functions studied, showed to be slightly better then the PDX-H2O models explored in this and previous contributions. For dimers all models are interchangeable, since all of them fit the ab initio curve with small errors; condensed phase systems are characterized by many-body interactions which account for the different performance of all models: for the systems studied, only the PD2-H2O model is able to catch all features of ion and water polarization, giving good results both for lithium–water and chloride–water systems. In fact, PDM-H2O and PD1-H2O highly underestimate water polarization such that in the dipole–dipole correlation plot all points are below the bisetrix. Such feature could not be observed in the calculations at gas phase because only the total dipole moment, rather than the molecular (ionic) one was computed; in that case it is highly probable that the ionic dipole moment overestimation could cancel out with the error in the water dipole moment. The good performance of PD2-H2O is probably due to its ability to describe the anisotropic polarizability of the water molecule [32].

For what concerns the parameterization of the damping functions, it has been shown that only the GAU-PD2-H2O model is able to reproduce all ab initio results both for ion and water polarization. This model is the only candidate for further force field development given its high portability from gas to condensed phase for the ions considered here.

Acknowledgments
This work was supported by FAR2007 of the Sassari University and by the Cybersar project managed by "Consorzio COSMOLAB".

[1] W. L. Jorgensen 2007 *J. Chem. Theory Comput.* **3** 1877.
[2] D. Horinek and R. R. Netz 2007 *Phys. Rev. Lett.* **99** 226104.
[3] J. Li, Z. Zhou and R. Sadus 2007 *J. Chem. Phys.* **127** 154509.
[4] S. Iuchi, S. Izvekov and G. A. Voth 2007 *J. Chem. Phys.* **126** 124505.
[5] J.-P. Piquemal, L. Perera, G. A. Cisneros, P. Ren, L. G. Pedersen and T. A. Darden 2006 *J. Chem. Phys.* **125** 054511.
[6] D. Jiao, C. King, a. Grossfield, T. A. Darden and P. Ren 2006 *J. Phys. Chem. B* **110** 18553.
[7] S. W. Rick and S. J. Stuart 2002 *Rev. Comp. Chem.* **18** 89.
[8] P. Jungwirth and D. J. Tobias 2006 *Chem. Rev.* **106** 1259.
[9] I-F. W. Kuo and C. J. Mundy 2004 *Science* **303** 658.
[10] *Curr. Opin. Colloid Interface Sci.* dedicated a special issue (Vol. 9, **2004**) reviewing the importance of salts in a wide range of chemical systems.
[11] M. G. Cacace, E. M. Landau and J. J. Ramsden 1997 *Rev. Biophys.* **30** 241.
[12] V. S. J. Craig, B. W. Ninham and R. M. Pashley 1993 *Nature* **364** 317.
[13] F. Lugli and F. Zerbetto 2007 *ChemPhysChem* **8** 47.
[14] D. Laage and J. T. Hynes 2007 *Proc. Natl. Acad. Sci USA* **104** 11167.
[15] B. Nigro, S. Re, D. Laage, R. Rey and James T Hynes 2006 *J. Phys. Chem. A* **110** 11237.
[16] M. Masia and R. Rey 2005 *J Chem Phys.* **122** 094502.
[17] A. W. Omta, M. F. Kropman, S. Woutersen and H. J. Bakker 2003 *Science* **301** 347.
[18] M. F. Kropman, H. K. Nyenhuys and H. J. Bakker 2002 *Phys. Rev. Lett.* **88** 77601.
[19] M. F. Kropman and H. J. Bakker 2001 *Science* **291** 2118.
[20] M. F. Kropman and H. J. Bakker 2001 *J. Chem. Phys.* **115** 8942.
[21] M. Masia, M. Probst and R. Rey 2004 *J. Chem. Phys.* **121** 7362.
[22] M. Masia, M. Probst and R.Rey 2005 *Comp. Phys. Comm.* **169** 331.
[23] M. Masia, M. Probst and R. Rey 2005 *J. Chem. Phys.* **123** 164505.
[24] M. Masia, M. Probst and R.Rey 2006 *Chem. Phys. Lett.* **420** 267.
[25] M. Masia 2008 *J. Chem. Phys.* **128** 184107.
[26] U. C. Singh and P. A. Kollman 1984 *J. Comp. Chem.* **5** 129.
[27] L. E. Chirlian and M. M. Francl 1987 *J. Comp. Chem.* **8** 894.
[28] C. M. Breneman and K. B. Wiberg 1990 *J. Comp. Chem.* **11** 361.
[29] B. H. Besler, K. M. Merz Jr. and P. A. Kollman 1990 *J. Comp. Chem.* **11** 431.
[30] G. A. Kaminski, H. A. Stern, B. J. Berne and R. A. Friesner 2004 *J. Phys. Chem. A* **108** 621.
[31] T. J. Giese and D. M. York 2005 *J. Chem. Phys.* **123** 164108.
[32] J.-P. Piquemal, R. Chelli, P. Procacci and N. Gresh 2007 *J. Phys. Chem. A* **111** 8170.
[33] P. Söderhjelm, A. Öhrn, U. Ryde and G. Karlström 2008 *J. Chem. Phys.* **128** 014102.
[34] M. J. Frisch et al. `Gaussian 03 Revision C.02`, Gaussian Inc., Wallingford CT, 2004.
[35] D. J. Tozer and N. C. Handy 1998 *J. Chem. Phys.* **109** 10180.
[36] R. Car and M. Parrinello 1985 *Phys. Rev. Lett.* **55** 2471.
[37] D. Marx and J. Hutter in *Modern Methods and Algorithms of Quantum Chemistry* 2000, NIC, FZ Jülich. For downloads see `http://www.theochem.rub.de/go/cprev.html`.
[38] A. D. Becke 1988 *Phys. Rev. A* **38** 3098.
[39] C. Lee, W. Yang and R. G. Parr 1988 *Phys. Rev. B* **37** 785.
[40] N. Troullier and J. L. Martins 1991 *Phys. Rev. B* **43** 1993.
[41] N. Marzari and D. Vanderbilt 1997 *Phys. Rev. B* **56** 12847.
[42] P. L. Silvestrelli 1999 *Phys. Rev. B* **59** 9703.
[43] G. Berghold, C. J. Mundy, A. H. Romero, J. Hutter and M. Parrinello 2000 *Phys. Rev. B* **61** 10040.
[44] J. Applequist, J. R. Carl and K. K. Fung 1972 *J. Am. Chem. Soc.* **94** 2952.
[45] W. H. Press, S. A. Teukolsky, W. T. Vetterling and B. P. Flannery in *Numerical Recipes in Fortran 77 - 2nd ed.*, 1992, Cambridge University Press, Cambridge.
[46] H. Coker 1976 *J. Phys. Chem.* **80** 2078.
[47] C. Hättig and B. A. Heß 1998 *J. Chem. Phys.* **108** 3863.
[48] P. Jungwirth and D. J. Tobias 2002 *J. Phys. Chem. A* **106** 379.
[49] A. Grossfield, P. Ren and J. W. Ponder 2003 *J. Am. Chem. Soc.* **125** 15671.
[50] B. T. Thole, 1981 *Chem. Phys.* **59** 341.
[51] I-F. W. Kuo, C. J. Mundy, B. L. Eggimann, M. J. McGrath, J. I. Siepmann, B. Chen, J. Vieceli and D. J. Tobias 2006 *J. Phys. Chem. B* **110** 3738.
[52] C. D. Wick, I-F. W. Kuo, C. J. Mundy and L. X. Dang 2007 J. Chem. Theory Comput. **3** 2002.
[53] M. Masia and R. Rey 2003 *J. Phys. Chem. B* **107** 2651.
[54] B. Chen, I. Ivanov, M. L. Klein and M. Parrinello 2003 *Phys. Rev. Lett.* **91** 215503.

Journal of Physics: Conference Series **177** (2009) 012005 doi:10.1088/1742-6596/177/1/012005

Quantum chemical study of water impact on the calcium hydroxyapatite

V. D. Khavryuchenko[1], O.V. Khavryuchenko[2], V.V. Lisnyak[2]

[1] Institute of Sorption and Problem of Endoecology, National Academy of Sciences of Ukraine, 13 General Naumov Str., UA-03167 Kyiv, Ukraine

[2] Kyiv National Taras Shevchenko University, 64 Volodymyrska Str., UA-01033, Kyiv, Ukraine

E-mail: alexk@univ.kiev.ua

Abstract. Structure and infra-red spectra of $Ca_5(PO_4)_3OH{\cdot}xH_2O$ (where $x = 1$–3) clusters were simulated by *ab initio* quantum chemical method. Evolution of the H-bonding character upon increase of water content and corresponding changes of the IR spectra in the ν(H–O) vibrations region are discussed.

Calcium hydroxyapatite ($Ca_5(PO_4)_3OH$) based composite materials (Ca–HAP) are frequently used in bone tissue engineering due to their high biocompatibility, cohesion and adhesion with metallic prosthesis [1–4]. Interaction of the Ca–HAP with water is of great scientific and practical interest due to changes of the material characteristics, which arise with complete or partial hydrolysis of calcium phosphates. Vibrational spectroscopy and, in particular, infra-red (IR) is an express-method to check the material's composition [1, 2, 5–9]. However, interpretation of the Ca–HAP IR spectra is a sophisticated task; since they can not be presented as the superposition of the minor phases' spectra. The interpretation problem can be solved via/by quantum chemistry (QC) modeling, which allows simulation of a spatial structure and IR spectra of the Ca–HAP materials.

Recently we have reported the results of spectroscopic and QC examination of the Ca–HAP clusters [10], including interaction of $Ca_3(PO_4)_2$ cluster with water molecules. It has been demonstrated that ν(H–O) band of the strongly hydrogen-bonded OH group of the $Ca_3(PO_4)_2{\cdot}xH_2O$ (where $x = 1$–3) clusters is shifted to lower frequencies up to 1000 cm^{-1} comparing to the one of the $Ca_5(PO_4)_3(OH)$ due to two-well potential [11] character of the bridging H atom. In fact, the $Ca_3(PO_4)_2$ undergoes hydrolysis, resulting in formation of acidic P–O–H and basic Ca–O–H groups. However, the acidic proton is closely bonded to both oxygen atoms of PO_4^{3-} group and water molecule.

The purpose of this communication is to present results of the QC study on ($Ca_5(PO_4)_3OH$) cluster interaction with water. Clusters $Ca_5(PO_4)_3OH{\cdot}xH_2O$ (where $x = 1$–3) have been simulated by *ab initio* methods [12, 13] in MINI basis set in the framework of PC–GAMESS software [12, 13], analogously to previously reported in [10]. The COSPECO software [14] has been used to simulate the vibrational spectra.

Two major aspects of the $Ca_5(PO_4)_3OH$ cluster interaction with water should be highlighted, namely i) evolution of the H-bonding character upon increase of water content and ii) corresponding changes of the IR spectra in the ν(H–O) vibrations region.

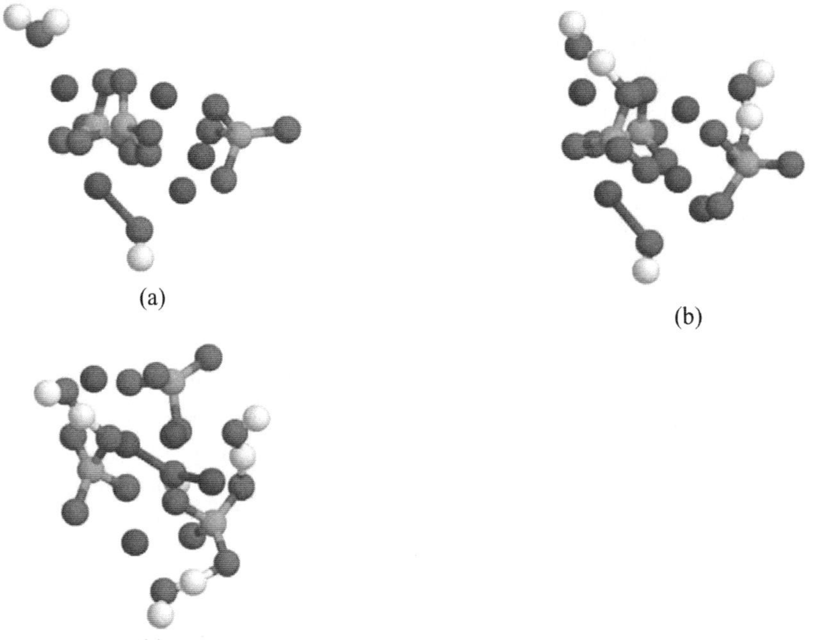

(a) (b) (c)

Figure 1. The QC optimized structures of the QC simulated clusters $Ca_5(PO_4)_3OH \cdot 1H_2O$ (a), $Ca_5(PO_4)_3OH \cdot 2H_2O$ (b) and $Ca_5(PO_4)_3OH \cdot 3H_2O$ (c). Oxygen atoms are black, hydrogens are white, phosphorus are light gray and calcium are dark gray.

The QC optimized structures of the simulated clusters are demonstrated on figure 1. As one can see, the water molecule in the cluster $Ca_5(PO_4)_3OH \cdot 1H_2O$ just weakly O-coordinates to the Ca^{2+} ion (Ca–O distance is 2.362 Å) without interaction with phosphate oxygens. However, in the cluster $Ca_5(PO_4)_3OH \cdot 2H_2O$ the water molecules form both coordination bonds to the Ca^{2+} ions and H-bonds to the phosphate oxygens. In the cluster $Ca_5(PO_4)_3OH \cdot 3H_2O$ the water molecules differ: there are two H_2O molecules, which coordinate both to Ca^{2+} and phosphate oxygens, as in the later case, and the third water molecule, forming stronger H-bond with the phosphate oxygen in two-well potential (O–H distances are 1.100 and 1.374 Å).

Placing the O–H distances in a series by ascending and analyzing, one can conclude that the strength of H-bonding of the third water molecule in the cluster $Ca_5(PO_4)_3OH \cdot 3H_2O$ is intermediate between the ones in the $Ca_3(PO_4)_2 \cdot 1H_2O$ and in $Ca_5(PO_4)_3OH \cdot 2H_2O$ clusters. The O–H distances in the former are extremely short – 1.198 Å and 1.215 Å, while in the later they are close to the standard H-bond – 1.468–1.471 Å and 1.010–1.051 Å. One should note that the computed distances of the H-bonds are probably understated due to cluster approach limitations (interaction of the water molecule with one phosphate groups is not compensated by attraction to the neighboring phosphate, increasing the distance).

Structural data on the level of water H-bonding in the $Ca_5(PO_4)_3OH \cdot xH_2O$ clusters series correlates with QC evaluated IR spectra well. The position of ν(H–O) bands in the cluster $Ca_5(PO_4)_3OH \cdot 1H_2O$ (Figure 2a) is situated in the high-frequency region and is typical for O-coordinated water. Appearance of the weakly H-bonded water molecule in the cluster $Ca_5(PO_4)_3OH \cdot 2H_2O$ results in the low-frequency shift of the ν(H–O) mode and, thus, splitting of the band (Figure 2b). Presence of the H_2O molecule, strongly H-bonded with the phosphate oxygen, in the cluster $Ca_5(PO_4)_3OH \cdot 3H_2O$ changes the picture drastically both in qualitative and quantitative way. A series of strongly low-shifted modes arise (Figure 2c), similarly to the case of the $Ca_3(PO_4)_2 \cdot 1H_2O$ spectrum [10]. Moreover, these modes are very mixed and can not be unambiguously assigned to a certain group vibrations due to high perturbation of the vibrations, arising from high level of interaction between vibrational coordinates. At the present state-of-art there are no conventional algorithms to obtain unique solution of this problem automatically, it only can be solved by manual selection of local force field for different parts of the studied system (i.e. by choice of negligible force constants), which is unrealistically time-consuming. Thus, one can only say that the bands in the 2000–3500 cm^{-1} region of the IR spectrum correspond to the interacting water-phosphate vibrations. (The scaling procedure [10] was not applied to these spectra since the bands can not be assigned uniquely, and, hence, the scaling factors can not be used correctly).

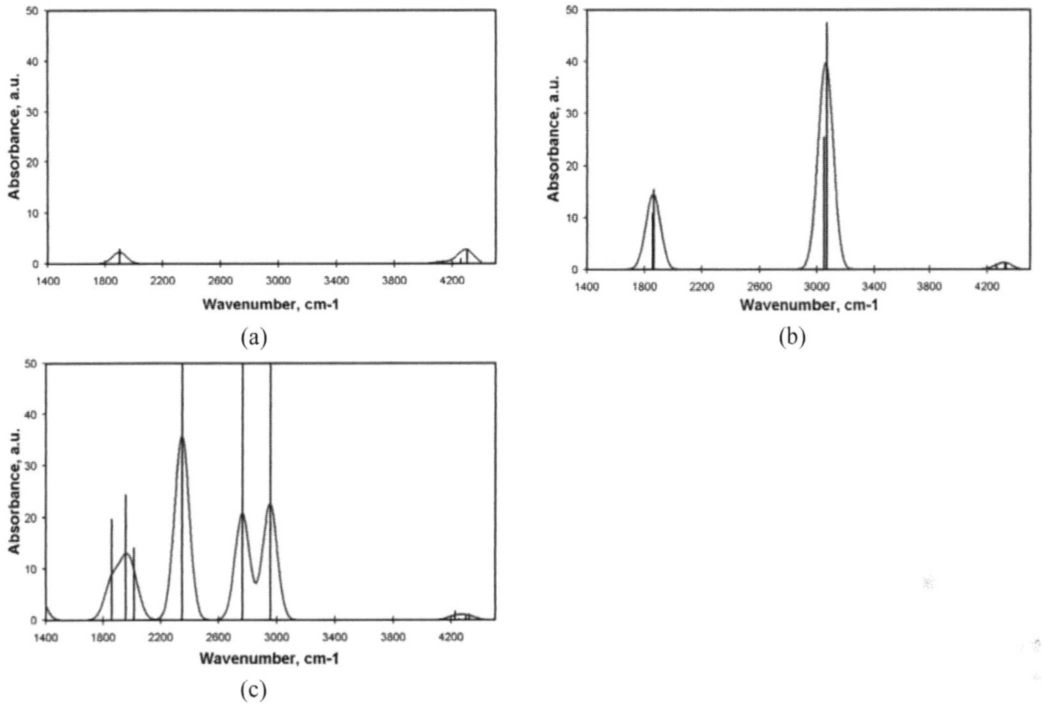

Figure 2. The QC simulated IR spectra of the clusters $Ca_5(PO_4)_3OH{\cdot}1H_2O$ (a), $Ca_5(PO_4)_3OH{\cdot}2H_2O$ (b) and $Ca_5(PO_4)_3OH{\cdot}3H_2O$ (c) in high-frequency region. Half-width constant $\sigma = 50$ cm^{-1}. (NB. The QC evaluated frequencies are systematically shifted to high-frequencies region.)

However, several important conclusions can be derived from the simulations data. In contrast to the $Ca_3(PO_4)_2$ hydrolysis, the interaction of the $Ca_5(PO_4)_3OH$ hydroxyapatite cluster with small quantities of water does not lead to severe structural changes, e.g. formation of new functional groups or dissociation of species. Thus, the IR spectra of hydroxyapatite samples with low content of water admixtures are convenient and can be interpreted as a superposition of $Ca_5(PO_4)_3OH$ units and weakly H-bonded H_2O. However, the hydroxyapatite samples with high content of water should have a broad intensive band at 3000–2000 cm^{-1} region.

References

[1] Vallet-Regí M and González-Calbet J M 2004 *Prog. Solid State Chem.* **32** 1
[2] Babini G N and Tampieri A 2004 *British Ceram. Trans.* **103** 101
[3] Bolognesi M P, Pietrobon R, Clifford P E and Vail T P 2004 *J. Bone Joint Surg.* A **86** 2720
[4] Dumbleton J and Manley M T *J.* 2004 *Bone Joint Surg.* A **86** 2526
[5] Bigi A, Boanini E, Gazzano M, Rubini K and Torricelli P 2004 *Bio-Med. Mater. Eng.* **14** 573
[6] Ignjatovic N and Uskokovic D 2004 *Spectroscopy* **18** 553

[7] Slosarczyk A and Piekarczyk J 1999 *Ceram. Int.* **25** 561
[8] Zeng H and Lancefield W R 2000 *Biomaterials* **21** 21
[9] Weinlaender M, Beumer J, Kenney E B and Moy P K 1992 *J. Mater. Sci.: Mater. in Medicine* **3** 397
[10] Khavryuchenko V D, Khavryuchenko O V and Lisnyak V V 2007 *J. Solid State Chem.* **180** 702
[11] Ratajczak H and Orville-Thomas W J 1980 *Molecular Interactions* (Chichester and NY: Wiley)
[12] Schmidt M W, Baldridge K K, Boatz J A, Elbert S T, Gordon M S, Jensen J H, Koseki S, Matsunaga N, Nguyen K A, Su S J *et al.* 1993 *J. Comput. Chem.* **14** 1347
[13] A. A. Granovsky, PC GAMESS version 7.1, http://classic.chem.msu.su/gran/gamess/index.html
[14] Khavryutchenko V D 2004 *Eurasian ChemTech. J.* **6** 157

Journal of Physics: Conference Series **177** (2009) 012006 doi:10.1088/1742-6596/177/1/012006

Neutron scattering studies of dynamic crossover phenomena in a coupled system of biopolymer and its hydration water

S H Chen[1,*], F Mallamace[1,2], X Q Chu[1], C Kim[1], M Lagi[1,4], A Faraone[2], E Fratini[4] and P Baglioni[4]

[1] Department of Nuclear Science and Engineering, Massachusetts Institute of Technology, Cambridge, MA 02139

[2] Dipartmento di Fisica and CNISM, Universita' di Messina, Vil. S. Agata CP 55, 98166 Messina, Italy

[3] NIST Center for Neutron Research, Gaithersburg, MD 20899-8562 and Department of Material Science and

[4] Department of Chemistry and CSGI, University of Florence, 50019 Italy

E-mail: sowhsin@mit.edu

Abstract. We have observed a Fragile-to-Strong Dynamic Crossover (FSC) phenomenon of the α-relaxation time and self-diffusion constant in hydration water of three biopolymers: lysozyme, B-DNA and RNA. The mean squared displacement (MSD) of hydrogen atoms is measured by Elastic Neutron Scattering (ENS) experiments. The α-relaxation time is measured by Quasi-Elastic Neutron Scattering (QENS) experiments and the self-diffusion constant by Nuclear Magnetic Resonance (NMR) experiments. We discuss the active role of the FSC of the hydration water in initiating the dynamic crossover phenomenon (so-called glass transition) in the biopolymer. The latter transition controls the flexibility of the biopolymer and sets the low temperature limit of its biofunctionality. Finally, we show an MD simulation of a realistic hydrated powder model of lysozyme and demonstrate the agreement of the MD simulation with the experimental data on the FSC phenomenon in the plot of logarithm of the α-relaxation time vs. 1/T.

1. Introduction

Water molecules in a protein solution may be classified into three categories: (i) the bound internal water; (ii) the surface water, i.e., the water molecules that interact with the protein surface strongly; and (iii) the bulk water. The bound internal water molecules, which occupy internal cavities and deep clefts, are extensively involved in the protein-solvent H-bonding and play a structural role in the folded protein itself. The surface water, which is usually called the hydration water, is approximately the first layer of water molecules that interacts with the solvent-exposed protein atoms of different chemical character, feels the topology and roughness of the protein surface, and exhibits the slow dynamics (a-relaxation). Finally, water that is not in direct contact with the protein surface but continuously exchanges with the surface water has properties approaching that of bulk water. The hydration water is believed to have an important role in controlling the biofunctionality of the protein. In this short note, we shall present some neutron scattering results of hydrated powder protein and two other hydrated biopolymers, B-DNA and RNA. In these cases, the hydration water represents the water in category (i) and (ii) mentioned above.

[*] To whom any correspondence should be addressed.

2. Role of Hydration Water in Glass Transition of Biopolymers

Functions of many globular proteins generally show sharp slowing-down around the temperatures between 200 and 240 K [1,2]. Experimental [3-5] and computational [6-8] results show a sharp increase of $< x^2 >$ of hydrogen atoms in proteins at about T_D = 220 K, which suggests that the dynamic transition (sometimes called the glass transition) may be occurring in the proteins at this temperature. There is strong evidence that this dynamic transition of protein is solvent-induced, since the hydration water of a protein also shows a kind of dynamic crossover at a similar temperature [9-11].

In our previous research [12], it was demonstrated that this dynamic transition of hydration water on lysozyme protein was, in fact, a FSC at 220 K. We also showed that the hydration water on B-DNA exhibits a similar FSC at T_L = 222 K [13]. More recently, we have studied a case of hydration water in RNA, which shows the FSC at a similar temperature, T_L = 220 K [14]. These results suggest that the dynamic transition temperature T_D in biopolymers happen at a universal temperature $T_D \approx 220 \pm 5$K, which is the consequence of the fact that it is triggered by the FSC of its hydration water, i.e., $T_D=T_L$. This remarkable fact is clearly shown in figure 1 and figure 2 for protein, DNA and RNA respectively, in spite of the difference in their chemical backbone structures.

3. Supportive Evidence from an MD Simulation of a Realistic Model of Hydrated Powder Lysozyme

A super-Arrhenius-to-Arrhenius dynamic crossover phenomenon has been observed in the translational α-relaxation time and in the inverse of the self-diffusion constant both experimentally and by simulations for lysozyme hydration water in the temperature range of $T_L = 223 \pm 2$ K [16]. MD simulations are based on a realistic hydrated powder model, which uses the TIP4P-Ew rigid molecular model for the hydration water. The convergence of neutron scattering, nuclear magnetic resonance and molecular dynamics simulations supports the interpretation that this crossover is a result of the gradual evolution of the structure of hydration water from a high-density liquid to a low-density liquid form upon crossing of the Widom line above the possible liquid-liquid critical point of water [15].

Figure 3 shows the MD simulation box containing two lysozyme protein molecules and 484 water molecules (hydration level h = 0.3 gram H_2O/gram dry protein). In order to simulate a realistic powder sample, the simulation box has to contain at least two protein molecules and its hydration water, because as one can see, the water molecules are not forming a uniform single layer surrounding a single protein molecule. Instead, there are fewer water molecules in between two proteins when they come into contact in close distance.

As one can see from figure 4a, the relaxing cage model (RCM) fits of the intermediate scattering function (ISF), both the time dependence and the Q-dependence, are excellent, allowing us to extract $< \tau_T >$ as a function of temperature, as shown in figure 4b. The crossover feature is clearly visible looking at the decay of the ISF below and above T_L. We fitted the fragile part with a VFT expression, $< \tau_T > = 5.0 \exp[300/(T-182)]$, and the strong part with an Arrhenius equation, $< \tau_T > = 5.0 \exp[1704/T]$. The crossover temperature is determined to be T_L = 221 K, very close to the experimental value of 220 K [12].

Acknowledgements

Research at MIT is supported by DE-FG02-90ER45429; at University of Florence by CSGI; at University of Messina by MURST-PRIN2004. We are grateful to NIST Centre for Neutron Research for allocating ample neutron beamtime to carry out the above experiments.

Journal of Physics: Conference Series **177** (2009) 012006 doi:10.1088/1742-6596/177/1/012006

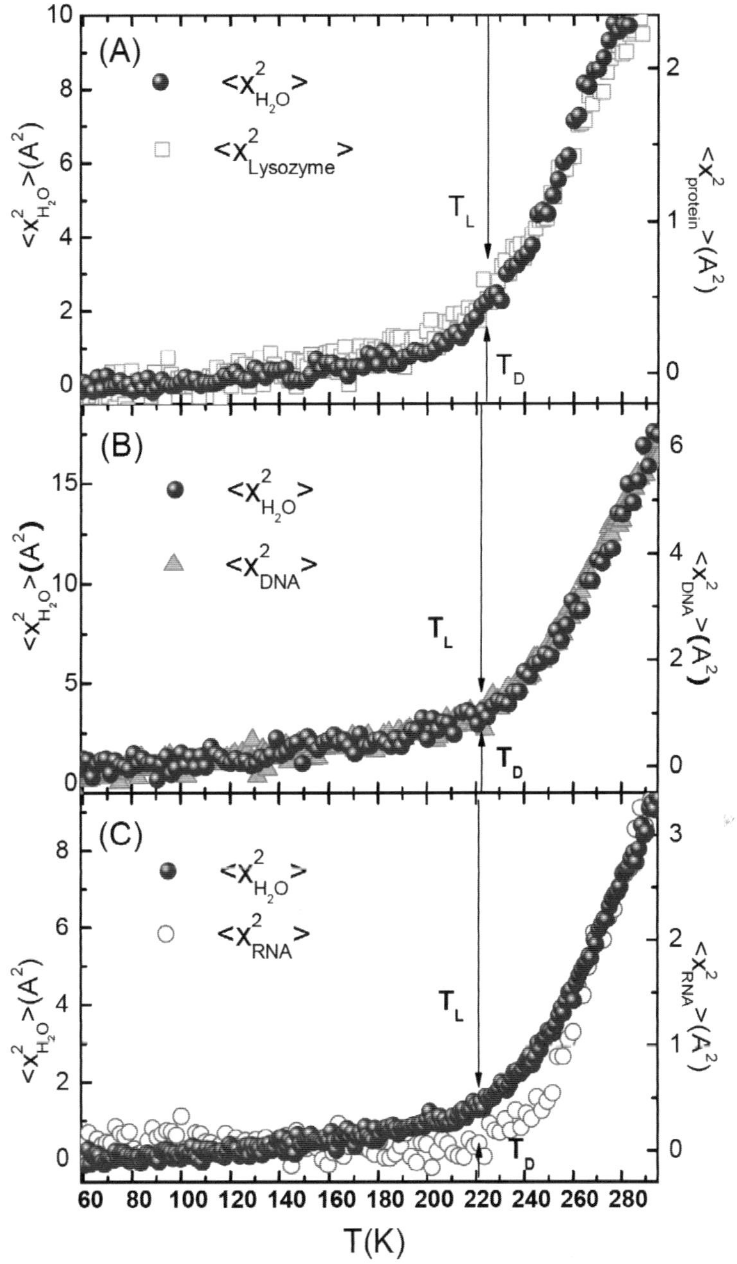

Figure 1. Each panel shows the temperature dependence of the MSD of hydrogen atoms in both the biopolymer and its hydration water, respectively. It shows evidence that the crossover temperatures of the two systems, the biopolymer and its hydration water, are closely synchronized. (A): MSD of hydrated lysozyme; (B): MSD of the hydrated B-DNA; (C): MSD of the hydrated RNA. The arrow signs indicate the approximate positions of the crossover temperature in both the biopolymer (T_D) and its hydration water (T_L). Note that the scale on the left hand side is for MSD of the hydration water and that on the right hand side is for the biopolymer.

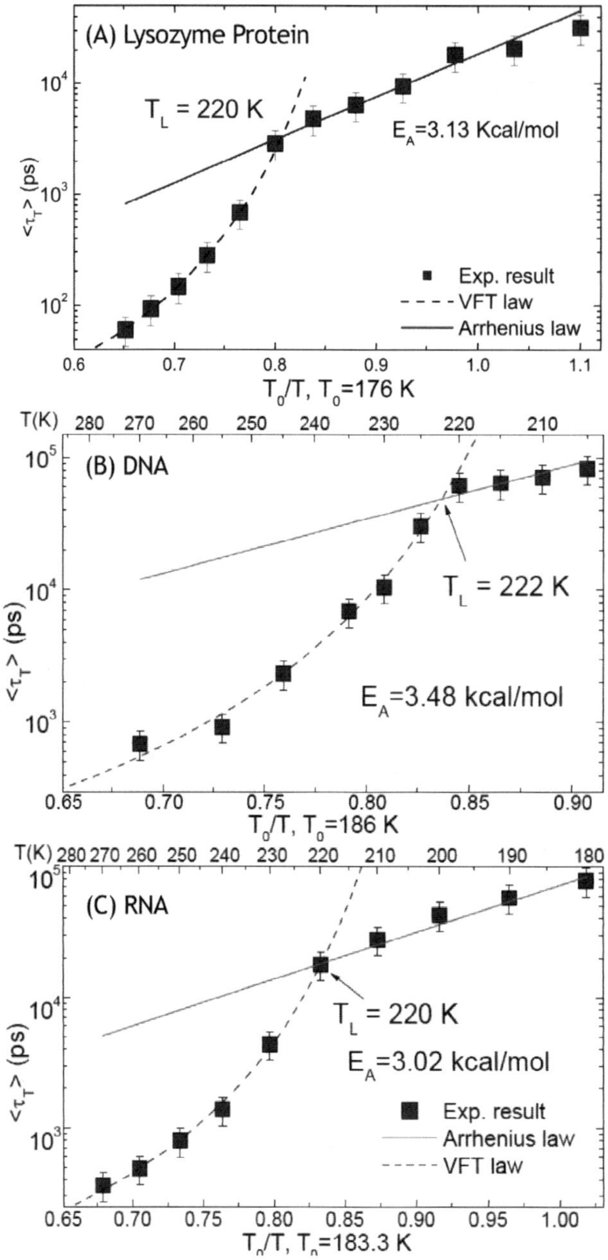

Figure 2. The extracted $< \tau_T >$ from fitting of QENS spectra by RCM plotted in the log scale *vs.* $1/T$. Panels (A), (B), and (C) show clearly well defined cusp-like dynamic crossover behavior in each case. The dashed line represents fitted curves using the VFT law, while the solid line is the fitting according to the Arrhenius law. FSC temperatures are respectively, (A) 220 K, hydration water in lysozyme; (B) 222 K, hydration water in DNA; and (C) 220 K, hydration water in RNA [12-14]. All three temperatures are essentially the same within the experimental error of 5 K suggesting universality of T_L in hydration water of all biopolymers.

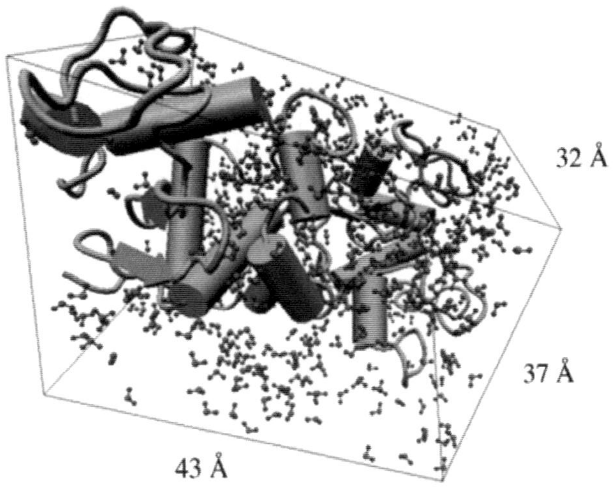

Figure 3. Snapshot from a MD simulation at T=250 K and P=1 atm of the hydrated lysozyme powder model, containing two protein molecules and 484 water molecules around them (hydration level h = 0.3) [16].

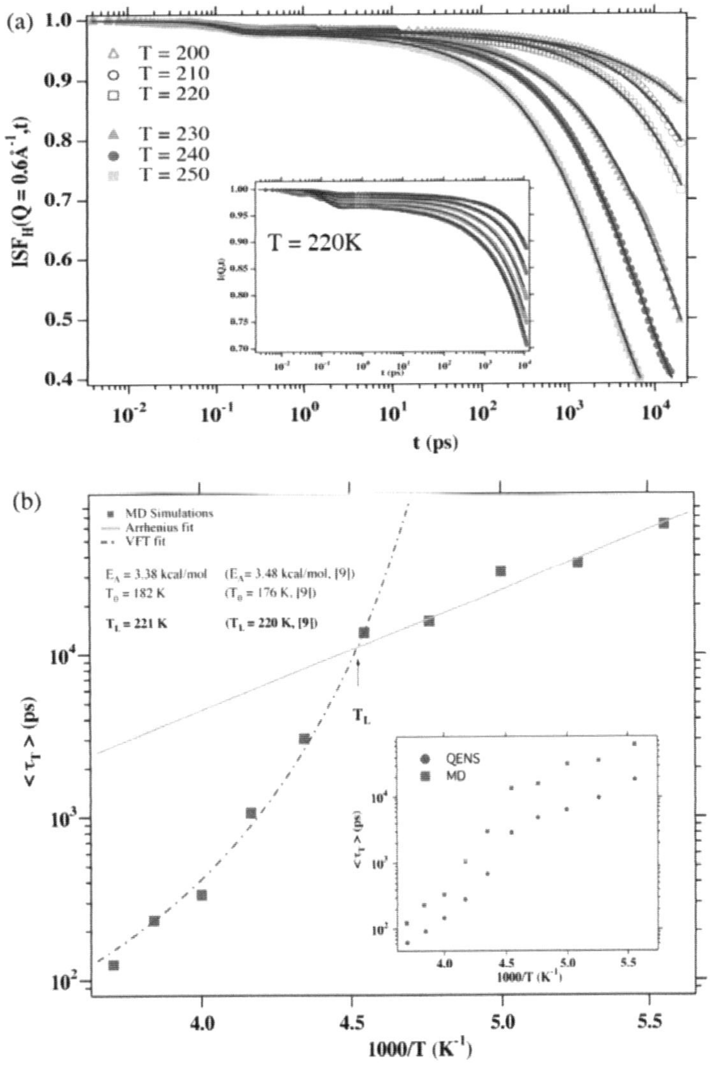

Figure 4. (a) Water proton incoherent self-intermediate scattering functions calculated at six different temperatures. (inset) ISF at five different Q-values (from top to bottom, 0.4, 0.5, 0.6, 0.7, and 0.8 $Å^{-1}$). The choice of the Q range was dictated by the low-limit value of Q = 0.2 $Å^{-1}$ imposed by the box dimensions and the high-limit value of Q = 1 $Å^{-1}$, below which rotational motions can be neglected. The solid curves are fits to the relaxing cage model in a wide time range of 7 orders of magnitude, between 2 fs and 20 ns. (b) Temperature dependence of the average translational relaxation time, $< \tau_T >$, from MD simulation; T_0 is the ideal glass transition temperature. Numerical data are fitted with a Vogel-Fulcher-Tammann (VFT) law at high temperatures and with an Arrhenius law at low temperatures (solid lines) but with the same prefactor [16].

References:

[1] Rasmussen B F, Stock A M, Ringe D and Petsko G A 1992 *Nature* **357** 423

[2] Daniel R M, Smith J C, Ferrand M, Hery S, Dunn R and Finney J L 1998 *Biophys. J.* **75** 2504

[3] Doster W, Cusack S and Petry W 1989 *Nature* **337** 754

[4] Ferrand M, Dianoux A J, Petry W and Zaccai G 1993 *Proc. Natl. Acad. Sci. USA* **90** 9668

[5] Tsai A M, Neumann D A and Bell L N 2000 *Biophys. J.* **79** 2728

[6] Tarek M and Tobias D J 2002 *Phys. Rev. Lett.* **88** 138101

[7] Tournier A L, Xu J and Smith J C 2003 *Biophys. J.* **85** 1871

[8] Kumar P, Yan Z, Xu L, Mazza M G, Buldyrev S V, Chen S H, Sastry S and Stanley H E 2006 *Phys. Rev. Lett.* **97** 177802

[9] Paciaroni A, Bizzarri A R and Cannistraro S 1999 *Phys. Rev. E* **60** R2476

[10] Caliskan G, Kisliuk A and Sokolov A P 2002 *J. Non-Crys. Sol.* **307-310** 868

[11] Fenimore P W, Frauenfelder H, McMahon B H and Parak F G 2002 *Proc. Natl. Acad. Sci. USA* **99** 16047

[12] Chen S H, Liu L, Fratini E, Baglioni P, Faraone A and Mamontov E 2006 *Proc. Natl. Acad. Sci. USA* **103** 9012

[13] Chen S H, Liu L, Chu X Q, Zhang Y, Fratini E, Baglioni P, Faraone A and Mamontov E *J. Chem. Phys.* 2006 **125** 171103

[14] Chu X Q, Fratini E, Baglioni P, Faraone A and Chen S H 2008 *Phys. Rev. E* **77** 011908

[15] Chu X Q, Faraone A, Kim C, Fratini E, Baglioni P, Leao J B and Chen S H "Pressure dependence of the dynamic crossover temperatures in protein and its hydration water", submitted to *J. Am. Chem. Soc.*

[16] Lagi M, Chu X Q, Kim C, Mallamace F, Baglioni P and Chen S H 2008 *J. Phys Chem B.* **112** 1571

Journal of Physics: Conference Series **177** (2009) 012007 doi:10.1088/1742-6596/177/1/012007

Looking for the best experimental conditions to detail the protein solvation shell in a binary aqueous solvent via Small Angle Scattering

Maria Grazia Ortore[1], Raffaele Sinibaldi[1], Francesco Spinozzi[1], Andrea Carbini[1], Flavio Carsughi[1,2] and Paolo Mariani[1]

1 Dipartimento di Scienze Applicate ai Sistemi Complessi e Unità CNISM, Università Politecnica delle Marche, Ancona, Italy

2 Institut für Festkörperforschung, Forschungszentrum, Jülich, Germany

E-mail: mg.ortore@alisf1.univpm.it

Abstract. Protein hydration features attract particular interest in different fields, from biology up to physics, crossing chemistry and medicine. Particular attention is devoted to proteins dissolved in binary aqueous mixtures, since the presence of cosolvent can induce modifications in structural and functional properties. We have recently developed a methodology to obtain a quantitative description on protein solvation shell by a set of in-solution small angle scattering experiments, simultaneously analysed by a global-fit approach. In this paper, numerical simulations of small angle scattering curves are presented to figure out the sensitivity of the technique to different experimental conditions. Simulations concern two model proteins of different molecular weights and an unique cosolvent. A reliability test is introduced in order to find the best experimental conditions to be investigated, together with the most suitable scattering probe (neutrons or X-rays).

1. Introduction

It is tremendously obvious that hydration, or, more generally speaking, solvation processes play an essential role in determining structural properties and stability of biological macromolecules in solution and in controlling their dynamical and functional properties. Indeed, water interfaces and water interplay with biological material likely retain important keys to understand many relevant biological processes. However, the whole picture is still largely incomplete from both biological and physical-chemical points-of-view, and this special issue on water interface demonstrates the peculiar interest devoted to the subject.

Protein hydration in water and protein preferential solvation in aqueous mixed solutions are two interesting cases. Considering the importance of this issue, it is pointless to enumerate the experimental approaches attempting to detail the protein hydration mechanisms or to describe the characteristics of the hydration shell and the consequent effects on protein stability. Concerning in particular the water in the first hydration shell, the debated question about its density was successfully faced in 1998 by Svergun and Zaccai, *via* a combination of Small Angle X-Ray and Neutron Scattering techniques [1] (SAXS and SANS, respectively; SAS in general). The setting of protein first shell density, in experimental conditions quite similar to those existing *in vivo*, is without doubts a milestone, although this result had been already predicted by molecular dynamic simulations [2] and crystallographic studies [3].

On the other side, the presence of a cosolvent in the aqueous solution can affect protein stability [4, 5], which can be either increased (see for example the case of glycerol addition in protein aqueous solutions) or reduced (as after the addition of urea). Moreover, the cosolvent can also play an important role in protein-protein association [6], even without modification of their structural characteristics. Pioneering studies related to cosolvent effects on proteins started more than 50 years ago [7, 8], but the issue is still much debated [5, 9, 10] nowadays. In particular, the discussion involves the two main different approaches used to investigate the possible outcome of cosolvent in protein aqueous solutions: the study of the solvent network modifications induced by cosolvents (and their related effect on proteins) [11, 12] and the study of the direct effect of cosolvents on protein surface and structure [13, 14].

Many efforts have been devoted to determine the protein solvation shell composition in the presence of a cosolvent by using different thermodynamic techniques, such as dialysis equilibrium and vapor pressure osmometry [15, 16, 17, 13, 18]. Some thermodynamic inconsistencies in a critique of "osmotic stress" were deeply discussed [9], as well as Kirkwood-Buff integrals and the preferential binding parameter, were recently derived to calculate various properties of infinitely dilute proteins in aqueous mixed solvents [19]. Finally, as cosolvent addition may lead either to protein aggregate formation or to an increased repulsion between particles, protein-protein interactions in the presence of a cosolvent have been also widely investigated [20, 21, 6, 22, 23, 24].

Therefore, the study of protein hydration and protein solvation in aqueous mixed solvent is still requiring new and effective experimental and theoretical efforts. This paper focus on a very recent approach, that we have developed to derive the solvation properties of a protein dissolved in water/cosolvent mixtures by in-solution SAXS and SANS experiments [22, 23]. The method is centered on a global and simultaneous analysis of all the measured SAS curves by using a thermodynamic model for the protein solvation, as the equilibrium exchange between water and cosolvent molecules from the solvation layer to the bulk described by Schellman [25]. Indeed, we have observed that the characteristics of the protein-solvent interface can be described if a sufficiently large number of different experimental conditions (*i.e.* protein concentration, composition of the aqueous mixed solvent, content of deuterated compounds) are considered. The application of SAS techniques should be commented: the main advantage is the possibility to perform the solvation analysis in a controlled environment, eventually quite similar to that observed *in-vivo*. In fact, the experiments concern proteins dissolved in a solution, at various protein concentrations and in buffer conditions that can approach the cellular environment; by contrast, all the experimental thermodynamic techniques previously considered can be applied mainly at infinite protein dilution. In addition, SAS enables the simultaneous investigation of protein solvation shell composition and of protein-protein interactions.

In this work, we will discuss how the successful application of this SAS "global fitting" strategy to derive quantitative information on the protein solvation shell is strongly related to the number of investigated experimental conditions and to the correct choice of these conditions. The paper is organized in a few sections that, other than summarizing previous results, focus on the transfer of the solvent-exchange model [25] to SAS data analysis. A series of numerical simulations have been performed in order to show a way to select the experimental conditions that, being more sensible to protein solvation features, are more suitable to be investigated by SAS. This point is very important, in particular when experiments are scheduled at neutron or synchrotron facilities in a unique and finite beam-time allocation.

2. In-solution SAS experiments

The macroscopical differential coherent scattering cross section for monodisperse and randomly oriented protein particles dissolved in a solvent is

$$I(Q) = n_p P(Q) S_{\text{eff}}(Q) \tag{1}$$

where n_p is the protein number density, $S_{\text{eff}}(Q)$ is the effective structure factor and $P(Q)$ is the protein averaged squared form factor. In this equation, Q is the modulus of the scattering vector ($Q = 4\pi \sin\theta/\lambda$, being 2θ the scattering angle and λ the X-rays or neutrons wavelength).

To study protein solvation properties in aqueous mixed solvents, the protein form factor in Eq. 1 should explicitly refer to the protein solvation shell, while $S_{\text{eff}}(Q)$, which is related to the particle-particle interactions, should be modeled taking into account any possible effect related to the presence of the cosolvent. For example, the cosolvent can change the dielectric properties of the solution, modify the local structure of the protein surface and/or change the amount of counter-ions bound to it [22, 23, 24]. In the next paragraph, we will then show as form and structure factors for the solvated protein can be modeled according to the solvent-exchange thermodynamic process described by Schellman and the protein-protein interaction potential.

3. Thermodynamic model applied to SAS curve analysis

The considered system consists of a globular protein dissolved in an aqueous solution containing a cosolvent. Therefore, its elements are protein (p), water (w) and cosolvent (o) molecules, whose number densities are n_p, n_w and n_o, respectively. A unique protein conformation is assumed, and the cosolvent molecular volume is thought to be larger than water molecular volume. In the following, we will refer to the binary solvent composition indicating either the water molar fraction in the solvent, $x_w = n_w/(n_w + n_o)$, or the cosolvent molar concentration, $C_o = n_o/N_A$ (N_A is Avogadro's number). On the other side, the protein concentration is defined by $c_p = n_p M_p/N_A$, being M_p the protein molecular weight.

According to Schellman's approach [5], the protein molecule is surrounded by a solvent region (the solvation layer), whose composition can be different from the one of the bulk solvent (see Fig. 1). Moreover, water and cosolvent molecules are not strongly bound to any specific binding site, so that the exchange between a cosolvent and a water molecule involves a few k_BT of free energy (k_B is Boltzmann's constant, T the absolute temperature). In this view, a large cosolvent molecule in contact with the protein would preserve its own conformation flexibility, so that multiple contacts with the protein can be considered rare and negligible events [26]. Finally, it is assumed that the number of protein contact sites, here indicated by m, does not depend on the solvent composition and can be simply considered as the number of water molecules that belong to the first hydration shell (h) when $x_w = 1$. It is well-established that the thickness of the first hydration shell around a protein is about 3 Å [5]; hence, m can be calculated by the ratio between the volume of the first hydration shell, V_h, and the volume of the water molecules in contact with the protein, $\nu_{w,h}$.

Combining all these assumptions, we can write the "mild" 1:1 cosolvent/water (O/W) exchange equilibrium process over a single unspecific contact site (P_s) and its thermodynamic constant as

$$\text{P}_s\text{O} + \text{W} \rightleftharpoons \text{P}_s\text{W} + \text{O} \qquad K = \frac{x_{w,h}}{1 - x_{w,h}} \frac{1 - x_{w,b}}{x_{w,b}} \frac{f_o}{f_w} \tag{2}$$

In this equation $x_{w,h}$ represents the probability that sites, on average, are occupied by water, $x_{w,b}$ is the water molar fraction in the bulk, and f_w and f_o are water and cosolvent activity coefficients in the bulk (in molar fraction units).

Small-angle scattering techniques are particularly suitable to investigate the chemical-physical properties of such a system. In fact, according to the SAS experimental low-resolution, bulk

Journal of Physics: Conference Series **177** (2009) 012007 doi:10.1088/1742-6596/177/1/012007

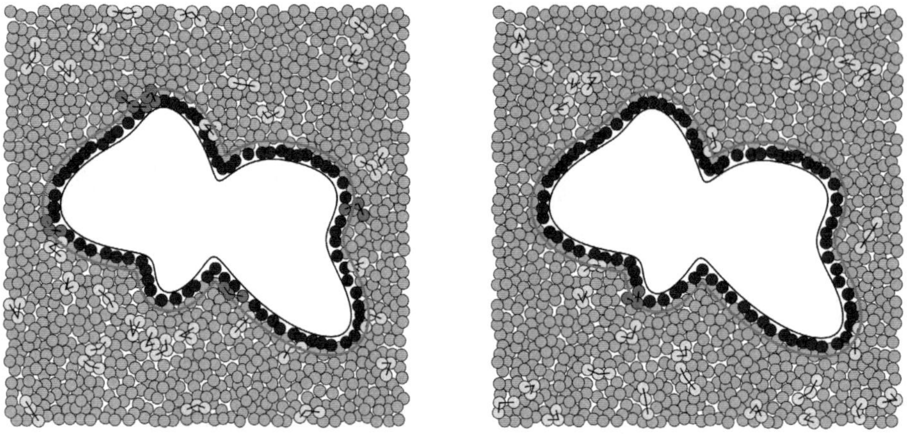

Figure 1. Scaled representations of a solvated BSA molecule based on PDB structure 1BKE, dissolved in a cosolvent/water mixture (C_o = 1 M) for $K = 1/2$ (left side) and $K = 2$ (right side). The light- and dark-blue circles represent water molecules (with diameter 3 Å) in the bulk and at contact with the protein, respectively. Three connected circles represent cosolvent molecules: those in the bulk are yellow, while those in contact with the protein are brown. The red line delimits the local domain shell with thickness $\delta_l = 4.2$ Å.

solvent and protein molecules can be considered as homogeneous scattering materials, each with its own scattering length density, ρ_b and ρ_p, respectively. In addition, a third region, between protein and bulk, may be visible by SAS if the protein is preferentially solvated ($K \neq 1$ in Eq. 2) and/or if solvent molecules in contact with protein undergo a variation of their volume. We refer to this region as *local domain* (l), whereas ρ_l is the corresponding scattering length density (see the red line in Fig. 1). Note that bulk solvent molecules fill the gaps among long cosolvent molecules hanged to the protein; therefore, both thickness and volume of local domain (δ_l and V_l, respectively) are larger than those of the first hydration shell and then, as a consequence, the local domain composition, $x_{w,l}$, is in-between $x_{w,h}$ and $x_{w,b}$. In particular, have we derived the following relations [22],

$$x_{w,l} = \begin{cases} \frac{x_{w,h}m\nu_o - x_{w,b}(m\nu_o - V_l) + x_{w,h}x_{w,b}(m\nu_{w,b} - V_h)}{x_{w,h}(m\nu_o - V_h) - x_{w,b}m(\nu_o - \nu_{w,b}) + V_l} & \text{for } x_{w,h} > \frac{m\nu_o - V_l}{m\nu_o - V_h} \\ x_{w,h} & \text{otherwise} \end{cases} \tag{3}$$

$$\nu_{w,l} = \frac{x_{w,h}(x_{w,b} - x_{w,l})}{x_{w,l}(x_{w,b} - x_{w,h})}(\nu_{w,h} - \nu_{w,b}) + \nu_{w,b} \tag{4}$$

In these equations, ν_o is the cosolvent molecular volume, that, according to the single-contact assumption, is considered to be the same both in the bulk and in the local domain, while $\nu_{w,b}$ and $\nu_{w,l}$ are the molecular volume of water in the bulk and in the local domains, respectively. Note that because electrostriction effects occurring at the protein surface, the molecular volume of the water in the first hydration shell can be quite different from the one of water in the bulk phase. Therefore, the water molecular volume in the local domain is included between $\nu_{w,b}$ and $\nu_{w,h}$, as indicated in Eq. 4. It should be also noticed that values of ν_o and $\nu_{w,b}$ can be found in literature as a function of $x_{w,b}$ for many water/cosolvent mixtures [27].

A further equation that, on the basis of protein number density (n_p) and core volume (V_p), relates the nominal solvent composition to the composition of the bulk and of the local domain, is reported here,

$$x_{w,b} = \frac{[1 - n_p(V_p + V_l)][x_{w,l}(\nu_o - \nu_{w,l}) - \nu_o]x_w - n_p V_l \nu_o (x_w - x_{w,l})}{[1 - n_p(V_p + V_l)][x_{w,l}(\nu_o - \nu_{w,l}) - \nu_o] + n_p V_l (\nu_{w,b} - \nu_o)(x_w - x_{w,l})}. \qquad (5)$$

Once the nominal system conditions have been fixed (n_p and x_w or, alternatively, c_p and C_o), when data or hypothesis regarding the dependence of activity coefficients f_w and f_o on the binary solvent composition are available, the set of Eqs. 2, 3, 4 and 5 can be solved and the bulk and local domain water molar fractions ($x_{w,i}$, $i = b, l$) can be calculated as a function of the exchange constant K.

The calculation of the domain scattering length densities is then straightforward,

$$\rho_i = \frac{x_{w,i}(a_w - a_o) + a_o}{x_{w,i}(\nu_{w,i} - \nu_{o,i}) + \nu_{o,i}} \quad \text{with } i = b, l \qquad (6)$$

$$\rho_p = \frac{a_p}{V_p} \qquad (7)$$

where a_j is the scattering length of the j molecule ($j = w, o, p$). For SAXS, $a_j = r_e(Z_j - z_j)$, being $r_e = 0.28 \cdot 10^{-12}$ cm the classical electron radius, Z_j the sum of atomic numbers of all atoms belonging to the j molecule and z_j its charge (in proton charge unit e). For SANS, a_j is the sum over all atoms of the atomic coherent neutron scattering lengths. It should be noticed that in this work, we considered SANS experiments performed on samples prepared using a binary solvent of deuteration grade $x_D \equiv n_{D+}/(n_{H+} + n_{D+})$, being n_{D+} and n_{H+} the number densities of all the exchangeable deuterons and protons present in the system, independently on the molecule they origin from. In this context, the scattering length of the j molecule results $a_j = a_{j,D}\, x_D + a_{j,H}\,(1 - x_D)$, where $a_{j,D}$ and $a_{j,H}$ are the sums of the atomic scattering lengths of all the atoms of the j molecule when all its exchangeable hydrogens are deuterons or protons, respectively [28, 29]. Notice that both $a_{j,D}$ and $a_{j,H}$ are modified when non-exchangeable hydrogens of the j molecules are substituted with deuteriums.

We are now able to write the averaged form factor, $P_1(Q)$, and the averaged squared form factor $P(Q)$ of the solvated protein [30, 31, 32]:

$$P_1(Q) = (\rho_p - \rho_b)V_p P_{1p}(Q) + (\rho_l - \rho_b)V_l P_{1l}(Q) \qquad (8)$$

and

$$P(Q) = (\rho_p - \rho_b)^2 V_p^2 P_{pp}(Q) + (\rho_l - \rho_b)^2 V_l^2 P_{ll}(Q) + 2(\rho_p - \rho_b)(\rho_l - \rho_b)V_p V_l P_{pl}(Q) \qquad (9)$$

where the single partial form factor $P_{1i}(Q)$ is the Fourier transform of the probability that a point at distance r from the protein center lies in the i domain ($i = p, l$), while the pair partial form factors $P_{ii'}(Q)$ ($i, i' = p, l$) are the Fourier transform of the probability to find a segment of length r with an end inside the i domain and the other end inside the i' domain.

The volumes V_i and the functions $P_{1i}(Q)$ and $P_{ii'}(Q)$ can be calculated with the Montecarlo method previously described [31, 32], considering the protein structure and the local domain thickness δ_l. Note that, being both $P_{1i}(Q)$ and $P_{ii'}(Q)$ dimensionless quantities ($P_{1i}(0) = P_{ii'}(0) \equiv 1$), if the two constrasts ($\rho_p - \rho_b$) and ($\rho_l - \rho_b$) are fixed, then the relative weight of the second term in Eq. 8, with respect to the first term, mainly depends on the particular ratio V_l/V_p of the investigated protein. We have also shown that, if the protein volume V_p changes by a factor γ close to 1 and if we assume subsequent isotropic changes of the protein shape, then the corresponding modifications of $P_{1i}(Q)$ and $P_{ii'}(Q)$ result in a simple scaling on the Q axis of a factor $Q\gamma^{1/3}$ [33].

As indicated by Eq. 1, SAS signal also depends on protein-protein interactions. To account for this effect, we need to calculate the structure factor $S(Q)$, which is the Fourier transform of the protein-protein correlation function $g(r)$. The physical origin of this correlation arises from the two-body interaction potential $u(r)$. Here, following Refs. [34, 35], we have written $u(r)$ as a combination of a classical hard sphere (HS) potential and two Yukawa terms, $u(r) = u_{\mathrm{HS}}(r) + u_D(r) + u_A(r)$. The first term defines the contact radius R of the protein,

$$u_{\mathrm{HS}}(r) = \begin{cases} \infty & r \leq 2R \\ 0 & r > 2R \end{cases} \tag{10}$$

while the general expression of the two Yukawa terms contains a constant B_j and an inverse decay length κ_j,

$$u_j(r) = B_j \frac{\exp[-\kappa_j(r-2R)]}{r} \qquad j = D, A \tag{11}$$

The $u_D(r)$ term represents the coulombic screened potential. Therefore, the constant B_D depends on the protein charge z_p and the bulk dielectric constant ε (which is a function of the bulk solvent composition $x_{w,b}$):

$$B_D = \frac{z_p^2 \mathrm{e}^2}{\varepsilon(1+\kappa_D R)^2}, \tag{12}$$

while κ_D is the inverse Debye screening length, which is a function of the ionic strength I_S of the solution (due to all microions, e.g. buffer, added salts and protein counterions):

$$\kappa_D = \left(\frac{8\pi \mathrm{e}^2 N_A}{k_B T \varepsilon} I_S \right)^{1/2} \tag{13}$$

The $u_A(r)$ Yukawa term represents an attractive potential, whose constant B_A can be written in terms of the value J of the potential at protein contact ($r = 2R$)

$$B_A = -2RJ \tag{14}$$

and the decay constant κ_A is a phenomenological constant.

The calculation of $S(Q)$ requires the solution of the Ornstein-Zernike (OZ) integral equations of the liquid state theory, within some approximate closure relation. Recently, Liu et al. [35] have proposed an efficient and robust solution in the mean spherical approximation (MSA), whose main result is the explanation for the formation of an extra low-Q peak in the $S(Q)$. Here, we are mainly interested in investigating the SAS signal at moderately low and high Q, where $P(Q)$ reflects differences in the local domain structures (see Figs. 2-3). Hence, we can follow the simpler approach of Ref. [34], i.e. the HS potential is the reference system, the two Yukawa terms are treated as a perturbation and the OZ is solved in the frame random phase approximation (RPA),

$$S(Q) = \frac{S_{\mathrm{HS}}(Q)}{1 + n_p S_{\mathrm{HS}}(Q)[U_C(Q) + U_A(Q)]/(k_B T)} \tag{15}$$

where $S_{\mathrm{HS}}(Q)$ is the solution of OZ under MSA of the HS potential

$$S_{\mathrm{HS}}(Q) = \left\{ 1 - \frac{12\eta[\eta(3-\eta^2)-2]}{(1-\eta)^4} \frac{j_1(2QR)}{2QR} \right\}^{-1} \tag{16}$$

being $\eta = \frac{4\pi}{3}R^3 n_p$ the volume fraction of the hard spheres, and $j_1(x)$ the 1st-order spherical Bessel function. $U_j(Q)$ is the Fourier transform of $u_j(r)$, which reads

$$U_j(Q) = 4\pi B_j \frac{\kappa_j \sin(2QR) + Q\cos(2QR)}{Q(Q^2 + \kappa_j^2)} \qquad j = D, A \tag{17}$$

Finally, the combination of form and structure factors gives the experimental SAS observable (see Eq. 1). It should be noticed that the so-called "effective" structure factor $S_{\text{eff}}(Q)$ is defined by

$$S_{\text{eff}}(Q) = 1 + \frac{[P_1(Q)]^2}{P(Q)}[S(Q) - 1] \tag{18}$$

but the ratio $[P_1(Q)]^2/P(Q)$ is different from 1 only when $QR \gg 1$: therefore, in the Q range usually investigated it can be considered $S_{\text{eff}}(Q) = S(Q)$.

4. Global Fit Strategy

In our terms, global fitting means the contemporary analysis of several scattering curves relative to different protein samples prepared at different solvent composition (*e.g.*, pH, ionic strength, cosolvent content) or measured at different conditions (*e.g.*, temperature, pressure, time) [6, 36, 22, 23]. If a good model for particle structure and interactions is used, and if the model is in some way sensitive to the experimental conditions, the procedure should be able to evidence even small effects on particle features (*e.g.*, hydration or solvation changes, conformational changes, aggregation state, particle and solvation shell densities). Indeed, we have shown that a global fit strategy can be successfully adopted to evaluate the efficiency of a theoretical model to reproduce all the SAS curves obtained from a wide set of different experimental conditions [6, 36].

Generally speaking, the whole scattering data are analyzed according to a unique model, considering a series of fitting parameters that are expected to be common to all samples (*e.g.*, protein structure factor, thickness of solvation shell, molecular volume of water in the bulk, molecular volume of the protein, association or dissociation constants in the case of aggregation equilibria) and a series of parameters which depend on solvent composition (*e.g.*, solvation shell composition, factors which describe the attractive potential, particle charge, ionic strength of the solution, solvent dielectric constant).

Here, we are considering the use of global fitting procedure to estimate the solvation characteristic of a protein dissolved in water in the presence of a cosolvent, according to the solvent exchange model described in the previous paragraph. The main common fitting parameter is then the equilibrium constant K, that describes the exchange equilibrium of water and cosolvent molecules between the protein solvation shell and the bulk solvent phase. As clearly expressed in Eq. 2, the K value determines water and cosolvent molar fractions in each domain and hence contributes to the solvated protein $P(Q)$ estimation. Other common fitting parameters are the molecular volume of water in the local domain, the local domain thickness and the protein volume. The protein structure can be described using the PDB crystallographic coordinates, but it should be clear that any structure modification, eventually induced by the cosolvent, has to be accurately investigated. Concerning the protein form factor, the model takes into account uniquely common parameters and the nominal composition of each investigated experimental condition.

More troublesome is the fit of parameters related to the structure factor. In fact, while the dielectric constant of the mixed solvent and the ionic strenght of solution, which play a role in the screened coulombic potential, are known, the protein charge and the attractive potential parameters, J and κ_A, require a particular attention. Even when the protein dissolved in the mixture maintains its structure, the surface charge can be modified: it is known that cosolvent molecules, even in small amounts in solution, can bind to specific sites of the protein backbone, locally reducing the barrier for exposing inner protein residues to the solvent, and hence changing the total protein surface charge [37, 38, 39]. In the choice of the fitting parameter, this issue cannot be neglected and the surface charge determination inside the global fitting procedure can be a direct way to assess local modifications induced by a rearrangement of cosolvent and water molecules around protein surface. Furthermore, cosolvents can also play a role in modifying attractive protein-protein potential [40]. Therefore, in some cases it can be very informative to consider interaction parameters as single curve fit parameters and maybe perform complementary experiments to verify the agreement between the obtained results.

5. Experimental set-up and Numerical Simulations

To perform a successful SAS global fitting experiment, it is necessary to assess for the best number of samples and the best sample conditions to be investigated. It is obvious in fact that the quality of results is related to the wideness of the experimental condition set, as the more scattering curves are globally fitted, the less uncertainty in the different common fitting parameters results. However, when SAS experiments are scheduled at neutron or synchrotron facilities in a unique and finite beam-time allocation, the number of samples that can be investigated is finite, and then the best experimental conditions should be preliminary established. Numerical simulations can be very useful to optimize number and experimental conditions on the basis of the available beam-time.

To elucidate this point, we refer to a "real" case, which concerns the analysis by SAS global fit of the preferential solvation of a protein dissolved in a mixed aqueous solution containing a cosolvent. To monitor eventual effects related to protein dimension, two different proteins have been considered, namely lysozyme, LYS, and bovine serum albumine, BSA. On the contrary, a unique cosolvent, with the molecular formula and the physics-chemical properties of urea, has been considered. However, the solvation properties (K) of the model cosolvent are considered unknown, so that it can be either preferentially excluded from or preferentially attached to the protein surface. Hence, the determination of the corresponding thermodynamic constant K, which can be larger, smaller or equal to unity, is the scope of the SAS experiment, that should be performed on a number of samples as smaller as possible.

In order to establish the best sample conditions from which the thermodynamic exchange constant can be derived by SAS global fit with the best accuracy, extended simulations of SAS curves in very different experimental conditions have been performed by using GENFIT software [41]. In particular, numerical simulations have been performed in the usually accessible Q-range from 0.03 to 0.4 Å^{-1}, considering the form factor in Eq. 9, using scattering length densities calculated according to Eqs. 6-7. Water molecular volume in the first hydration shell has been considered to be $\nu_{w,h} = 27.3$ Å^3 (10% less than the bulk water volume), because of electrostriction effects at the protein surface [1]. For the sake of simplicity, the activity coefficients f_w and f_o have been fixed to unity. The protein and local domain volumes (V_p and V_l), as well as the partial form factors $P_{ii'}(Q)$, have been calculated on the basis of the protein data bank (PDB) structure and considering a thickness of the local domain $\delta_l = 4.2$ Å. In the case of LYS the entry 6LYZ [42] was used, while for BSA, whose cristalline structure is not yet available, the entry 1BKE [43, 44], corresponding to human serum albumine, was used. We obtained $V_p = 17,200$ Å^3 for LYS and $81,000$ Å^3 for BSA, in agreement with literature [1, 23]. To model the structure factor, both protein charge and attractive potential parameters were derived from literature data. Assuming that the solvent pH is 7, we considered LYS and BSA charges as $z_p = 5$ and $z_p = 15$, respectively [45, 46]; the ionic strenght of the solutions was considered to be 50 mM and the dielectric constant to vary according to the cosolvent presence in solution [27]. Moreover, the temperature was fixed to $T = 298$ K. Attractive term parameters were fixed to $J = 3k_BT$ and $\kappa_A = 0.2$ Å^{-1}, in agreement with previous results [34, 22, 23].

It should be finally noticed that all simulations were repeated considering three different K values of $1/2, 1, 2$, which correspond to a protein local domain enriched, indifferent or pauperized in cosolvent with respect to the bulk. In this way, the set of simulated curves can be analyzed in searching for such experimental conditions (namely, C_o, c_p and x_D, globally indicated by the symbol i) in which the SAS curves, corresponding to different values of K, are significantly different.

6. Numerical simulation results

The whole sets of SAXS and SANS simulated curves are reported in Figs. 2 and 3. For both LYS and BSA, we calculated 3 *ensembles* of curves: SAXS from hydrogenated samples and SANS from samples at two deuteration grades x_D 1 and 0.8 ($x_\text{D} = 1$ maximizes the contrast between protein and water; $x_\text{D} = 0.8$ is the condition where the contrast between water and protein and between the considered cosolvent and water are equal [28]). Each ensemble consists of 5 different protein concentrations c_p (10, 40, 70, 100 and 130 g/l, *e.g.* from very dilute to highly concentrate conditions), each one at 5 different cosolvent molar concentrations C_o (0.5, 1, 1.5, 2, 2.5 M), resulting in 25 experimental condition. As for each experimental condition i, three different K values were considered ($K = 1/2, 1, 2$), the number of curves simulated for each ensemble is 75. Data in Figs. 2 and 3 show that SAS profiles vary for different K values, and that the differences are mostly pronounced in certain conditions (concentrations, deuteration grade, type of experiment). Note that such variations are especially evident in the lowest and highest Q ranges, as the local domain features reflect on the effective particle shape and the local domain thickness affects the short-range distances probability, respectively.

For each set i, to quantitatively characterize the difference between the two curves with $K = 1/2$ and $K = 2$ and the "reference" curve $I_1(Q)$ corresponding to $K = 1$, we have introduced the *resemblance factor* $R_f^{(i)}$ [47],

$$R_f^{(i)} = R_{f,1/2} + R_{f,2} \tag{19}$$

$$R_{f,K} = \int_{Q_\text{min}}^{Q_\text{max}} dQ \left\{ \frac{I_K(Q) - I_1(Q)}{[I_K(Q_\text{min})]^{1-\alpha}[I_K(Q)]^\alpha} \right\}^2 \tag{20}$$

where Q_min and Q_max are the minimum and the maximum value of the chosen experimental Q-range and $I_K(Q)$ is the macroscopical differential scattering cross section (Eq. 1) relative to the K value. In the assumption that errors over the SAS detector counts follow a Poisson statistics, it can be found that the errors of $I(Q)$ can be expressed as proportional to $I(Q)^\alpha$, with $\alpha \approx 0.2 - 0.8$, depending on experimental set-up parameters (transmissions, measurement times, etc.). Then, according to the standard χ^2 definition [48], we have chosen the $R_{f,K}$ definition reported in Eq. 20. Here, we fixed $\alpha = 0.5$. According to this definition, $R_f^{(i)}$ offers a quantitative estimation of how much the experimental condition i is able to evidence a protein local domain composition different from the bulk one: the bigger is $R_f^{(i)}$, the more convenient will be the i-condition. Values of $R_f^{(i)}$ calculated for the 25 i-conditions of each ensemble are reported in Tab. 1 for BSA and in Tab. 2 for LYS. It is evident that $R_f^{(i)}$ increases with both $c_{p,i}$ and $C_{o,i}$.

The first remarkable result concerns the large $R_f^{(i)}$ observed in SANS simulations with $x_\text{D} = 0.8$, proving that sample deuteration can improve SAS sensibility to different K values. However, further simulations (not shown) indicated that this effect is not linear, so that maximizing the sample deuteration grade does not correspond to maximize $R_f^{(i)}$. Such a result strongly confirms the importance of performing numerical simulations before defining the experimental conditions to be investigated.

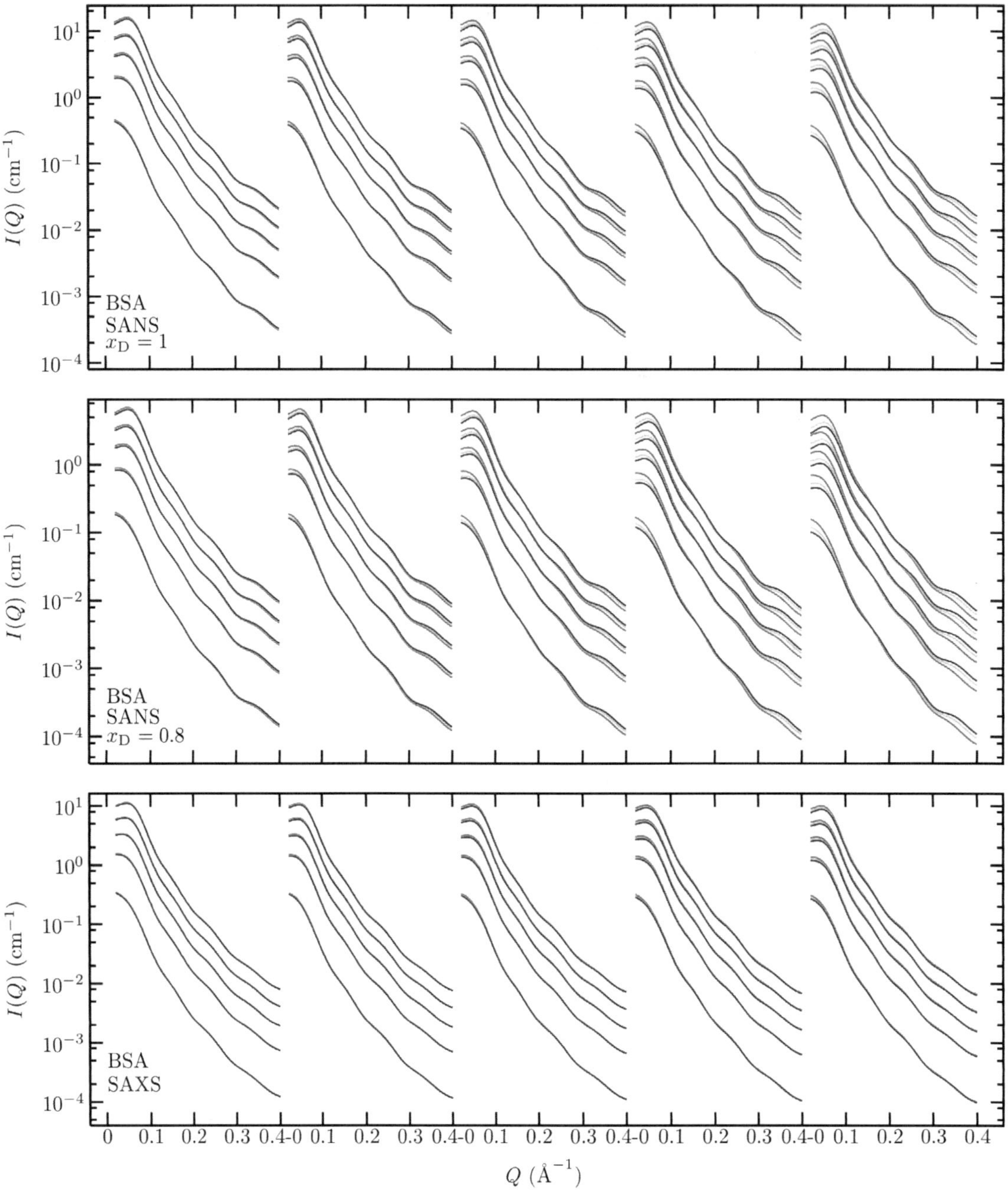

Figure 2. SANS (at $x_D = 1$ and 0.8) and SAXS numerical simulations of BSA. In each panel we report results obtained in water/cosolvent mixtures (from left to right: C_o = 0.5, 1, 1.5, 2 and 2.5 M) at different protein concentrations (from bottom to up: c_p = 10, 40, 70, 100 and 130 g L^{-1}). Blue, red and green lines refer to K = 1/2, 1 and 2, respectively. For the sake of clarity, in each panel, curves at the same c_p have been scaled upwards by a factor multiple of 1.5.

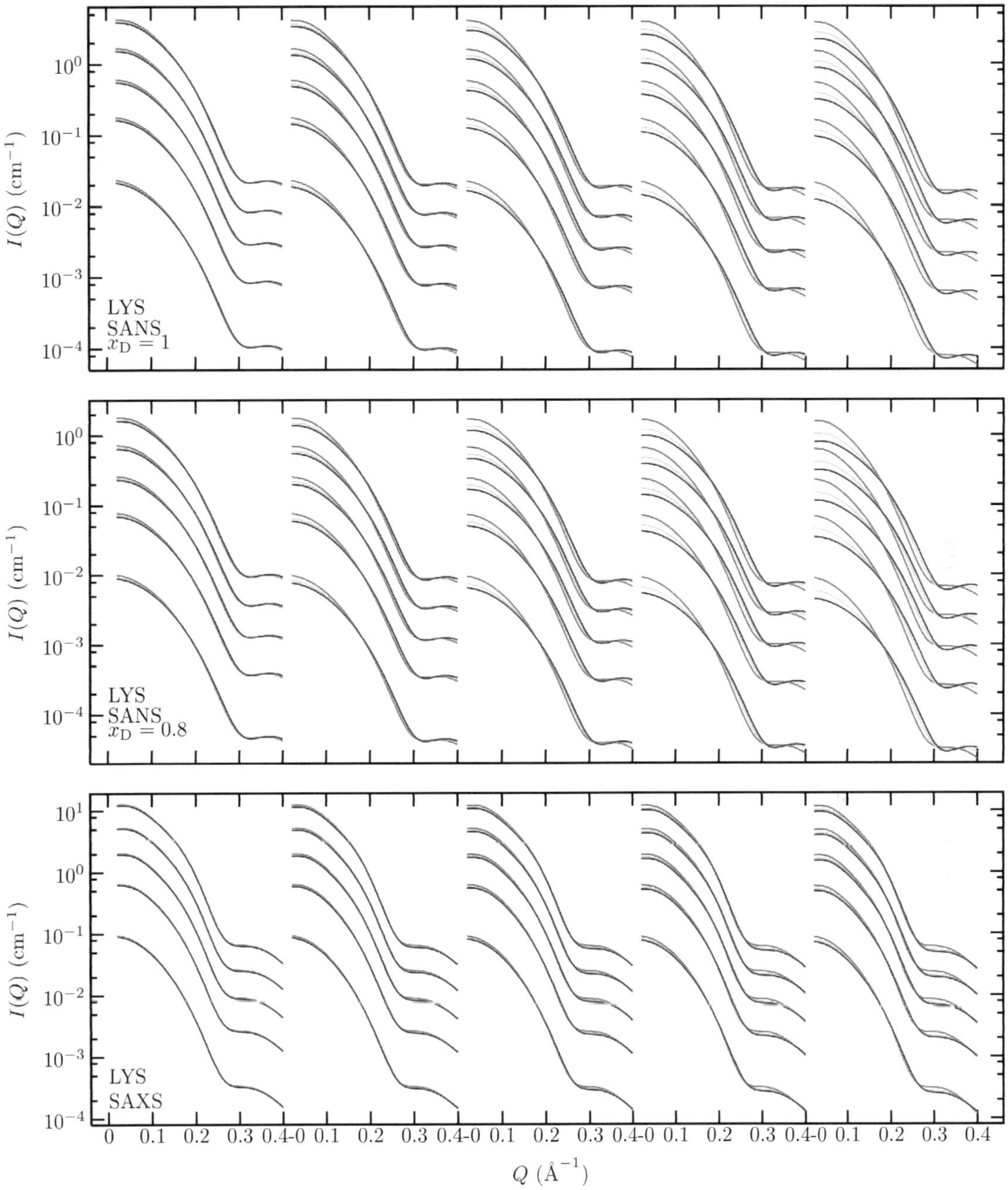

Figure 3. SANS (at $x_D = 1$ and 0.8) and SAXS numerical simulations of LYS. All other details as in Fig. 2.

c_p (g L^{-1})	C_o (M) 0.5	1	1.5	2	2.5
	$R_f^{(i)}$ (10^{-3} Å^{-1})				
	BSA SANS x_D=1				
10	0.0663	0.269	0.616	1.12	1.82
40	0.0802	0.325	0.747	1.37	2.21
70	0.0920	0.374	0.860	1.57	2.55
100	0.102	0.417	0.96	1.76	2.86
130	0.112	0.455	1.05	1.93	3.14
	BSA SANS x_D=0.8				
10	0.0934	0.382	0.888	1.65	2.72
40	0.113	0.463	1.08	2.00	3.31
70	0.130	0.532	1.24	2.31	3.82
100	0.144	0.593	1.38	2.58	4.28
130	0.158	0.649	1.52	2.83	4.71
	BSA SAXS				
10	0.0112	0.0454	0.104	0.187	0.295
40	0.0137	0.0557	0.127	0.230	0.363
70	0.0159	0.0647	0.148	0.268	0.424
100	0.0179	0.0731	0.167	0.303	0.480
130	0.0198	0.0809	0.186	0.336	0.533

Table 1. Resemblance factors $R_f^{(i)}$ calculated with Eq. 19 and relative to the three ensembles of numerical simulations of BSA reported in Fig. 2. Colored symbols on the right of each factor identify different sets j inside each ensemble. According to a "Union Jack" flag scheme, the different sets are color-coded and defined as: column , row , backslash , star , slash .

As it is also important to verify the applicability of the thermodynamic model (in this case, the Schellman solvent-exchange equilibrium model) and the accuracy of the protein-protein interaction pattern over a range of protein and cosolvent concentrations as wider as possible, a second factor related to the sampling size and design has been introduced. This factor has been called *concentration-distribution* factor $R_c^{(j)}$

$$R_c^{(j)} = \frac{2}{N_j(N_j-1)} \sum_{i>i' \in j}^{N_j} \left\{ \left[\frac{c_{p,i} - c_{p,i'}}{c_{p,\max} - c_{p,\min}} \right]^2 + \left[\frac{C_{o,i} - C_{o,i'}}{C_{o,\max} - C_{o,\min}} \right]^2 \right\}^{1/2}, \qquad (21)$$

where N_j is the number of experimental i conditions investigated, $c_{p,i}$ and $C_{o,i}$ are protein and cosolvent concentrations relative to the i-th condition, respectively. The protein concentration range $(c_{p,\max} - c_{p,\min})$ and the cosolvent concentration range $(C_{o,\max} - C_{o,\min})$ have to be wisely chosen by the experimentalist, from one hand to avoid protein aggregation and/or unfolding, and from the other hand to obtain a satisfactory signal/noise ratio. Notice that the more homogeneously distributed $c_{p,i}$ and $C_{o,i}$ are within their corresponding ranges, the higher is the factor $R_c^{(j)}$.

As an indicative example, we have chosen 5 different sets j within each ensemble, each one consisting of $N_j = 5$ i-conditions selected from those reported in Tabs. 1 and 2 following a "Union Jack" flag scheme. The different colors in the Tables indicate the different sets. For each set j, we have calculated both $R_c^{(j)}$ and the sum of the corresponding resemblance factors,

c_p (g L^{-1})	C_o (M) 0.5	1	1.5	2	2.5
	$R_f^{(i)}$ (10^{-3} Å^{-1})				
LYS SANS x_D=1					
10	0.312	1.24	2.81	5.08	8.21
40	0.321	1.28	2.89	5.24	8.47
70	0.329	1.31	2.97	5.39	8.71
100	0.337	1.34	3.05	5.52	8.95
130	0.344	1.37	3.12	5.65	9.17
LYS SANS x_D=0.8					
10	0.445	1.79	4.10	7.56	12.6
40	0.458	1.84	4.22	7.80	13.0
70	0.470	1.89	4.34	8.02	13.4
100	0.481	1.94	4.46	8.23	13.6
130	0.491	1.98	4.55	8.43	14.1
LYS SAXS					
10	0.0435	0.176	0.397	0.710	1.11
40	0.0486	0.196	0.445	0.796	1.25
70	0.0525	0.212	0.483	0.866	1.36
100	0.0557	0.226	0.516	0.925	1.46
130	0.0587	0.238	0.544	0.98	1.55

Table 2. Resemblance factors $R_f^{(i)}$ calculated with Eq. 19 and relative to the three ensembles of numerical simulations of LYS reported in Fig. 3. See Tab. 1 for all the other details.

$\sum_{i\in j} R_f^{(i)}$ (i.e. the sum of $R_f^{(i)}$ values marked with the same color in each table). Results are plotted in Fig. 4, left panels.

Looking over the $\sum_{i\in j} R_f^{(i)}$, it is straightforward to note that for each protein and for each set j, the best experiment for evidencing differences of composition between bulk and protein local domain is SANS performed on samples at deuteration grade $x_D = 0.8$ (green point-line). On the contrary, SAXS is the worse experiment (blue point-line), as already evidenced by the bottom curves of Figs. 2 and 3, which do not present remarkable differences for different K values. Instead, it can be considered unexpected that both SANS and SAXS numerical simulations for the smaller protein, LYS, show the higher sensibility (i.e. higher $\sum_{i\in j} R_f^{(i)}$ values) to different Ks with respect to the larger protein, BSA. However, it can be easily shown that, at a fixed thickness of the local domain δ_l, the ratio V_l/V_p decreases by increasing V_p: hence, as hinted by Eq. 9, the same variation of local domain composition (i.e. the same difference $\rho_l - \rho_b$) leads to a larger relative variation of $P(Q)$ for small proteins with respect to large proteins.

It is also interesting to note that $\sum_{i\in j} R_f^{(i)}$ presents different maxima for the different proteins (Fig. 4, left side). In the BSA case, the maximum is observed for the backslash set, while for LYS the maximum is observed when the star set is considered. Nevertheless, for both proteins, the two higher values are observed for backslash and star sets, and their difference is small.

However, as the experimental condition choice should also account for the test of applicability of both the thermodynamic and interaction models, the behavior of $R_c^{(j)}$ factor should be considered: for both proteins, the maximum is observed for the star sets. To derive a final parameter, we then take the mathematical product of $\sum_{i\in j} R_f^{(i)}$ and $R_c^{(j)}$ factors as a combined *reliability test*. Calculated values are reported in Fig. 4 (right side): in any case, and if only

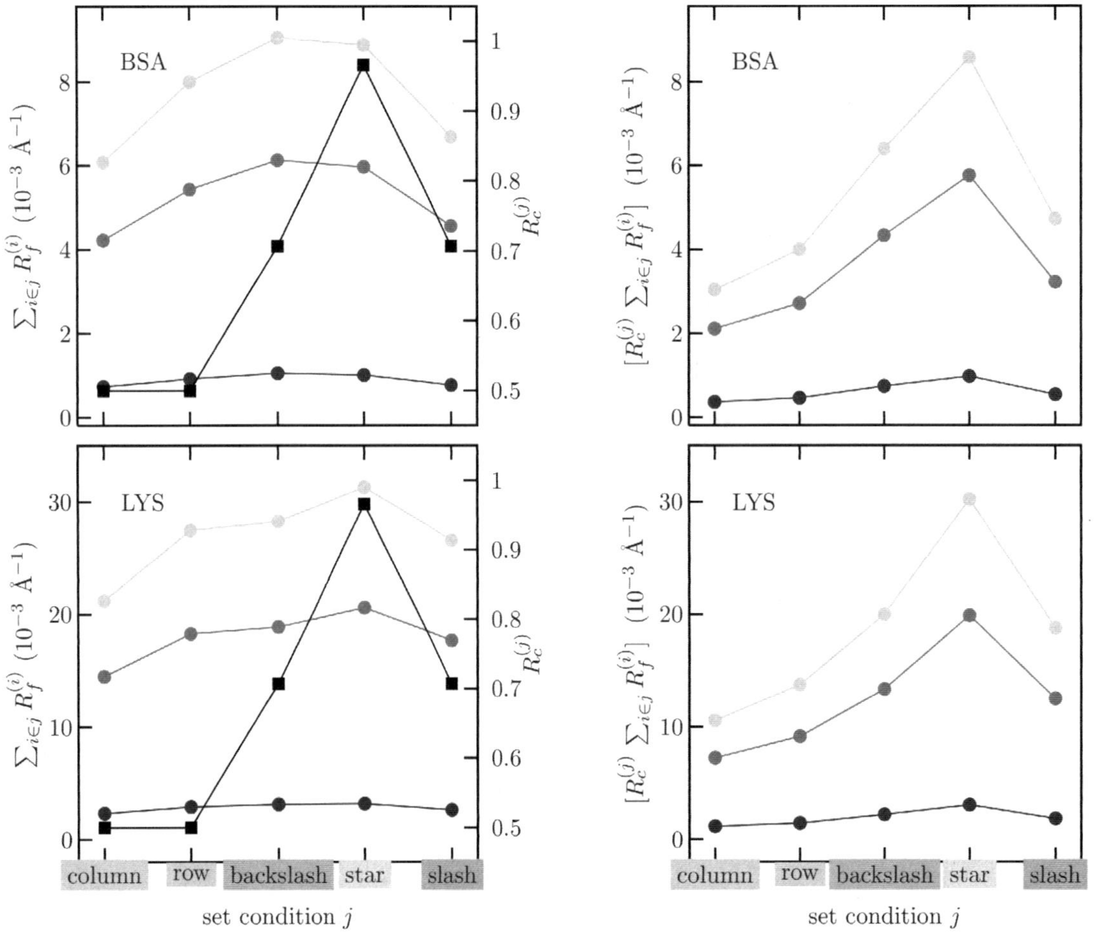

Figure 4. Left panels. Values of $\sum_{i\in j} R_f^{(i)}$ obtained according to the different sets j of experimental conditions (column , row , backslash , star , slash) as evidenced by colors in Tabs. 1 and 2. In the panels different colored point-lines correspond to SANS experiments at $x_D = 0.8$ (geeen), at $x_D = 1$ (red) and to SAXS experiments (blue). Black point-lines are concentration-distribution factors $R_c^{(j)}$ obtained for each set j of samples. Right panel. Mathematical products of $\sum_{i\in j} R_f^{(i)}$ and of $R_c^{(j)}$, obtained for each set j. Colors meaning as in left panels.

5 experiments can be performed, the best experimental choice concerns SANS experiments performed on samples prepared at $x_D = 0.8$ with protein and cosolvent concentrations defined by the star set. It should be observed that the obtained results can be different if a larger number of experiments can be scheduled or when the concentration limits (both for protein and cosolvent) are different: however, a good choice of experimental conditions can result in SAS curve differences that can be significantly very large (in the presented case, the resemblance factor changes from 6 to 10 times when the considered set moves from column to star).

7. Experimental Results

We briefly describe two successful applications of the methodology presented in Sec. 3 for both LYS and BSA. LYS was investigated both in water-glycerol [22] and in water-urea mixtures [24] by SANS, whereas BSA was investigated in the presence of urea by both SAXS and SANS techniques [23]. Glycerol and urea are two well-known and largely studied cosolvents. However, the molecular mechanisms that govern the stabilyzing and destabilyzing effects of these cosolvents are still debated. According to literature, stabilizing cosolvents are preferentially excluded from protein solvation shell, while the denaturing chemical agents are preferentially driven to protein surface [49, 50, 51]. Nonetheless, the quantitative estimation of cosolvent molecules located at protein surface was obtained at infinite protein dilution, or in a crystal. This is the reason why our method can be regarded as efficient in determining the existence of a thermodynamic exchange equilibrium in a wide range of protein concentrations and in experimental conditions closer to those *in-vivo*.

In the first study [22], we have determined the experimental conditions to be investigated via numerical simulations of the form factor of lysozyme in water-glycerol mixtures, considering the protein solvation shell enriched in water. 35 different experimental conditions (protein concentration, water/glycerol molar fraction in the solvent and deuteration grades) have been then experimentally investigated by SANS and analyzed using GENFIT software [41]. A thermodynamic constant $K = 1.87 \pm 0.03$ has been obtained, confirming that lysozyme dissolved in water/glycerol mixed solvents is preferentially hydrated. This result has been compared with literature data concerning the preferential binding coefficient in the case of Ribonuclease A [52, 19], evidencing a good agreement.

In the second study [23], the thermodynamic exchange constant has been determined by SAXS and SANS experiments on BSA dissolved in water-urea solutions. According to extended simulations, 9 SANS curves and 8 SAXS curves, corresponding to different protein and urea concentrations, were simultaneously fitted using GENFIT software [41]. The common parameter K resulted 0.6 ± 0.1, indicating that the protein solvation shell is enriched in urea. In this case, the BSA charge has been detected to slightly increase with addition of urea in solution, suggesting a denaturation mechanism in which the main process could be urea localization on the surface of the protein, producing local structural modifications (see also [24]).

It should be finally noticed that a direct relation between the preferential binding coefficients and the thermodynamic constant K was derived (see Eqs. 14-15 in [22]) and that the agreement with results arising from thermodynamic techniques [15, 49, 53, 52, 54, 55, 56] was very satisfactory.

8. Conclusions

In a binary aqueous solvent, the description of protein solvation shell, both in terms of thickness and composition, can be obtained by SAS experiments considering a convenient combination of experimental conditions. This paper offers the detailed guideline to select the type of experiment (SANS or SAXS) and the set of sample conditions to be investigated, in order to obtain successful results using a global fit analysis. As a thermodynamic relation is necessary to connect the investigated samples, the application of Schellman solvent exchange model to the global fit analysis of SAS experimental curves, has been widely described. Extended SAS curve simulations have been then presented and a reliability test to establish the best sample conditions introduced.

Therefore, this paper can be considered a sort of recipe that points the way to perform SAS experiments when the expected result is the determination of a quantitative protein solvation shell description, together with the estimation of protein-protein interactions parameters. Despite previous and successful approaches via thermodynamic techniques [15, 17, 13], we stress the relevance of this method because in-solution SAS concerns experimental environments even close to the *in-vivo* ones and includes the analysis of protein-protein interactions. Spreading of this methodology can lead to face other noticeable issues, like the effect of cosolvents on protein thermal stability [57], activity and aggregation.

9. Acknowledgments
MGO thanks European Science Foundation for financial support to participate to ESF-FWF conference "Water Interfaces in Physics, Chemistry and Biology: A Multi-Disciplinary Approach", Obergurgl, Austria, 8-13 December 2007.

Appendix A. List of symbols

Symbol	Description
α	exponent of the power-law of errors on $I(Q)$
a_j	scattering length of the j molecule ($j = w, o, p$)
$a_{j,\mathrm{D}}$	sum of the atomic scattering lengths of all the atoms of the j molecule ($j = w, o, p$) when all its exchangeable hydrogens are deuterons
$a_{j,\mathrm{H}}$	sum of the atomic scattering lengths of all the atoms of the j molecule ($j = w, o, p$) when all its exchangeable hydrogens are protons
B_j	constant of the j Yukawa potential ($j = D, A$)
C_o	nominal cosolvent molar concentration of the binary solvent
c_p	nominal protein mass concentration of the three-component solution
δ_l	local domain thickness
e	proton charge
ε	bulk dielectric constant
f_o	bulk cosolvent activity coefficient
f_w	bulk water activity coefficient
$g(r)$	protein-protein radial pair correlation function
γ	ratio between the core protein volume V_p and its nominal value
$I(Q)$	macroscopical differential scattering cross section
$I_K(Q)$	macroscopical differential scattering cross section of a system with exchange constant K
J	attractive potential at the contact
K	cosolvent/water exchange equilibrium constant
κ_A	inverse decay length of the attractive Yukawa potential
κ_D	inverse Debye screening length
m	number of protein contact sites
M_p	protein molecular weight
η	volume fraction of the hard spheres
N_A	Avogadro's number
n_{D^+}	nominal number concentration of all the exchangeable deuterons
n_{H^+}	nominal number concentration of all the exchangeable protons
N_j	number of conditions of the j set
n_o	nominal cosolvent number concentration of the three-component solution
n_p	nominal protein number concentration of the three-component solution
n_w	nominal water number concentration of the three-component solution
$P(Q)$	average squared form factor of the solvated protein
$P_1(Q)$	average form factor of the solvated protein
$P_{1i}(Q)$	single partial form factor of the i domain ($i = p, l$)
$P_{ii'}(Q)$	pair partial form factors of the i, i' domains ($i, i' = p, l$)
Q	scattering vector modulus
ρ_b	scattering length density of the bulk binary solvent
ρ_l	scattering length density of the local domain
ρ_p	scattering length density of the core protein
R	protein-protein contact radius
r_e	classical electron radius
$R_c^{(j)}$	concentration-distribution factor of the j set of experimental conditions
$R_f^{(i)}$	resemblance factor of the i set of curves

Symbol	Description
$R_{f,K}$	resemblance factor of a curve associated with the exchange constant K
$S(Q)$	protein-protein structure factor
$S_{\text{eff}}(Q)$	"effective" structure factor
$S_{\text{HS}}(Q)$	structure factor of the hard sphere potential in the mean spherical approximation
$u(r)$	protein-protein interaction potential
$u_A(r)$	attractive Yukawa potential
$u_D(r)$	coulombic screened Yukawa potential
$u_{\text{HS}}(r)$	hard sphere potential
$U_j(Q)$	Fourier transform of $u_j(r)$ $(j = D, A)$
V_h	first protein hydration shell volume
V_l	local domain volume
V_p	core protein volume
ν_o	molecular volume of the cosolvent
$\nu_{w,b}$	molecular volume of the water in the bulk binary solvent
$\nu_{w,h}$	molecular volume of water in contact with the protein
$\nu_{w,l}$	molecular volume of the water in the local domain
x_{D}	deuteration grade
x_w	nominal water molar fraction of the binary solvent
$x_{w,b}$	water molar fraction of the bulk binary solvent
$x_{w,h}$	average probability of water protein sites occupation
$x_{w,l}$	water molar fraction of the local domain
Z_j	sum of atomic numbers of all atoms belonging to the j molecule $(j = w, o, p)$
z_j	charge of the j molecule $(j = w, o, p)$ in proton charge unit e

[1] Svergun D, Richard S, Koch M H J, Sayers Z, Kuprin S and Zaccai G 1998 *Proc. Natl. Acad. Sci. USA* **95** 2267–2272
[2] Levitt M and Sharon R 1988 *Proc. Natl. Acad. Sci.* **85** 7557–7561
[3] Burling F T, Weis W I, Flaherty K M and Brunger A T 1996 *Science* **271** 72–77
[4] Klibanov A M 2001 *Nature* **409** 241–246
[5] Schellmann J A 2003 *Biophys. J.* **85** 108–125
[6] Ortore M G, Spinozzi F, Carsughi C, Mariani P, Bonetti M and Onori G 2005 *Chem. Phys. Lett.* **418** 338–342
[7] Frensdorff H K, Watson M T and Kauzmann W 1953 *J. Am. Chem. Soc.* **75** 5167–5172
[8] Kirkwood J G and Goldberg R J 1950 *J. Chem. Phys.* **18** 54–57
[9] Timasheff S N 2002 *Proc. Natl. Acad. Sci.* **99** 9721–9726
[10] Purkiss A, Skoilakis S and Goodfellow J M 2001 *Phyl. Trans. R. Soc. Lond. A* **359** 1515–1527
[11] Soper A K, Castner E W and Luzar A 2003 *Biophys. Chem.* **105** 649–666
[12] Stumpe M C and Grubmuller H 2007 *J. Phys. Chem. B* **111** 6220–6228
[13] RE Karen S T and Bloomfield V A 2002 *Biophys. J.* **82** 2876–2891
[14] Pike A and Acharya K R 1994 *Protein Sci.* **3** 706–710
[15] Zhang W, Capp M W, Bond J P, Anderson C F and Jr M T R 1996 *Biochem.* **35** 10506–10516
[16] Casassa E F and Eisenberg H 1964 *Adv. Protein Chem.* **19** 287–395
[17] Noelken M E and Timasheff S N 1967 *J. Biol. Chem.* **242** 5080–5085
[18] Smith P E 2006 *Biophys. J.* **91** 849–856
[19] Shulgin I L and Ruckenstein E 2005 *J. Chem. Phys.* **123** 054909
[20] Javid N, Vogtt K, Krywka C, Tolan M and Winter R 2007 *Chem. Phys. Chem.* **8** 679–689
[21] Niebuhr M and Koch M H J 2005 *Biophys. J.* **89** 1978–1983
[22] Sinibaldi R, Ortore M G, Spinozzi F, Carsughi F, Frielinghaus H, Cinelli S, Onori G and Mariani P 2007 *J. Chem. Phys.* **126** 235101–9
[23] Sinibaldi R, Ortore M G, Spinozzi F, de Souza Funari S, Teixeira J and Mariani P 2008 *Eur. Biophys. J.* 10.1007/s00249–008–0306–z
[24] Ortore M, Sinibaldi R, Spinozzi F, Carsughi F, Clemens D, Bonincontro A and Mariani P 2008 **submitted**
[25] Schellman J A 2005 *Quat. Rev. Biophys.* **38** 351–361
[26] Baynes B M and Trout B L 2003 *J. Phys. Chem. B* **107** 14058–14067
[27] Lide D R 1996 *Handbook of Chemistry and Physics, 77th ed.* (Cleveland, Ohio: CRC Press)
[28] Jacrot B 1976 *Reg. Prog. Phys.* **39** 911

[29] Dianoux A J and Lander G 2003 *Neutron Data Booklet* (Grenoble, France: OCP Science)
[30] Feigin L A and Svergun D I 1987 *Structure analysis by small-angle X-ray, neutron scattering.* (New York, Plenum Press)
[31] Spinozzi F, Carsughi F, Mariani P, Teixeira C V and Amaral L Q 2000 *J. Appl. Cryst.* **33** 556–559
[32] Spinozzi F, Gazzillo D, Giacometti A, Mariani P and Carsughi F 2002 *Biophys. J.* **82** 2165–2175
[33] Spinozzi F, Carsughi F, Mariani P, Saturni L, Bernstorff S, Cinelli S and Onori G 2007 *J. Phys. Chem. B* **111** 3822 –3830
[34] Narayanan J and Liu X Y 2003 *Biophys. J.* **84** 523–532
[35] Liu Y, Chen W R and Chen S H 2005 *J. Chem. Phys.* **122** 044507–13
[36] Spinozzi F, Carsughi F, Mariani P, Saturni L, Bernstorff S, Cinelli S and Onori G 2007 *J. Phys. Chem. B* **111** 3822 –3830
[37] Caballero-Herrera A, Nordstrand K, Berndt K D and Nilsson L 2005 *Biophys. J.* **89** 842–857
[38] Pike A C W and Acharya K R 1994 *Protein Sci.* **3** 706–710
[39] Bhattacharya J, Ghoshmoulick R, Choudhuri U, Chakrabarty P, Bhattacharya P K, Lahiri P, Chakraborti B and Dasgupta A K 2004 *Anal. Chim. Acta* **522** 207–214
[40] Farnum M and Zukoski C 1999 *Biophys. J.* **76** 2716–2726
[41] Spinozzi F 2008 *http://alisf1.univpm.it/biophysics/software.htm*
[42] Diamond R 1974 *J. Mol. Biol.* **82** 371–391
[43] Curry S, Mandelkow H, Brick P and Franks N 1998 *Nat. Struct. Biol.* **5** 827–835
[44] Svergun D I, Petoukhov M V and Koch M H J 2001 *Biophys. J.* **80** 2946–2953
[45] Tanford C and Buzzel J G 1956 *J. Phys. Chem.* **60** 225–231
[46] Tanford C and Roxby R 1972 *Biochem.* **11** 2192–2198
[47] Carbini A *Private communication*
[48] Press W H, Teukolsky S A, Vetterling W T and Flannery B P 1994 *Numerical Recipies. The Art of Scientific Computing* (Cambridge, U.K.: Cambridge University Press)
[49] Timasheff S N 1998 *Adv. Protein Chem.* **51** 355–432
[50] Timasheff S N and Xie G 2003 *Biophys. Chem.* **105** 421–448
[51] Shimizu S and Smith D J 2004 *J. Chem. Phys.* **121** 1148
[52] Gekko K and Timasheff S N 1981 *Biochem.* **20** 4667–4676
[53] Shimizu S 2004 *J. Chem. Phys.* **120** 4989–4990
[54] Xie G and Timasheff S N 1997 *Biophys. Chem.* **64** 25
[55] Shulgin I L and Ruckenstein E 2006 *Biophys. J.* **90** 704–707
[56] Shimizu S and Matubayasi N 2006 *Chem. Phys. Lett.* **420** 518–522
[57] Spinozzi F, Ortore M G, Sinibaldi R, Mariani P, Esposito A, Cinelli S and Onori G 2008 **submitted**

Journal of Physics: Conference Series **177** (2009) 012008 doi:10.1088/1742-6596/177/1/012008

Influence of the water molecules near surface of viral protein on virus activation process

Shepelenko S.O.[1,2]**, Salnikov A.S.**[1]**, Rak S.V.**[1]**, Goncharova E.P.**[1]**, Ryzhikov A.B.**[1]

[1] Federal State Research Institution State Research Center of Virology and Biotechnology VECTOR of the Federal Service for Surveillance in Consumer Rights Protection and Human Well-being (FSRI SRC VB VECTOR)
Koltsovo, Novosibirsk Region, Russia

[2] Novosibirsk State University, Novosibirsk, Russia

E-mail: shep@vector.nsc.ru; shep@ngs.ru.

Abstract. The infection of a cell with influenza virus comprises the stages of receptor binding to the cell membrane, endocytosis of virus particle, and fusion of the virus envelope and cell endosome membrane, which is determined by the conformational changes in hemagglutinin, a virus envelope protein, caused by pH decrease within the endosome. The pH value that induces conformation rearrangements of hemagglutinin molecule considerably varies for different influenza virus strains, first and foremost, due to the differences in amino acid structure of the corresponding proteins. The main goal of this study was to construct a model making it possible to assess the critical pH value characterizing the fusogenic activity of influenza virus hemagglutinin from the data on hemagglutinin structure and experimental verification of this model. Under this model, we assume that when the electrostatic force between interacting hemagglutinin molecules in the virus envelop exceeds a certain value, the hemagglutinin HA1 subunits are arranged so that they form a cavity sufficient for penetration of water molecules. This event leads to an irreversible hydration of the inner fragments of hemagglutinin molecule in a trimer and to the completion of conformational changes. The geometry of electrostatic field in hemagglutinin trimer was calculated taking into account the polarization effects near the interface of two dielectrics, aqueous medium and protein macromolecule. The critical pH values for the conformational changes in hemagglutinin were measured by the erythrocyte hemolysis induced by influenza virus particles when decreasing pH. The critical pH value conditionally separating the pH range into the regions with and without the conformational changes was calculated for several influenza virus H1N1 and H3N2 strains based on the data on the amino acid structure of the corresponding hemagglutinin molecules. Comparison of the theoretical and experimental values of critical pH values for influenza virus strains suggests that the proposed model of the interaction between water molecules and influenza virus envelope proteins has a high prediction efficiency.

Introduction

The life cycle of the majority of viruses is a series of successive stages of virion interaction with the sensitive cell, namely, attachment of the virus particle to the cell surface; penetration into the cell; release of the virus genome; transcription, translation, and processing of the virus proteins; assembly of daughter virions; and release of virus from the infected cell. The infection of a cell with influenza virus commences from the interaction between virus glycoprotein, hemagglutinin, and the terminal sialooligosaccharides of cell membrane proteins. Such receptor-type binding to the membrane induces the endocytosis of the virus particle by the host cell, which is accompanied by the inflow of hydrolases to the endosome. Activation of hydrolases by acidification of the endosome medium leads to degradation of the virus particle. In turn, a decrease in pH also activates virus envelope proteins inducing conformational molecular rearrangement in their structure, which leads to juxtaposition of the virus and cell endosome membranes and their fusion. Consequently, a pore is formed between the virus and the cytoplasm, through which the virus genome enters the cell [1] [2]. Thus, the biological activity of a virus is determined by the ability of virus hemagglutinin to rearrange the conformation of its molecule in a pH-dependent manner and to induce the fusion of virus and endosome membrane before the virus proteins are degraded by hydrolases upon pH decrease in the endosome.

The monomer of hemagglutinin is a prolate globule located on the envelope perpendicular to the virion surface. The monomer consists of two domains, HA1 and HA2, connected with a disulfide bond located at a distance of 115 A from the molecule's top and of 20 A from the lipid bilayer envelope of the virus [3]. Hemagglutinin forms a quaternary structure comprising three monomers with their HA1 domains turned outside and HA2 domains concealed inside.

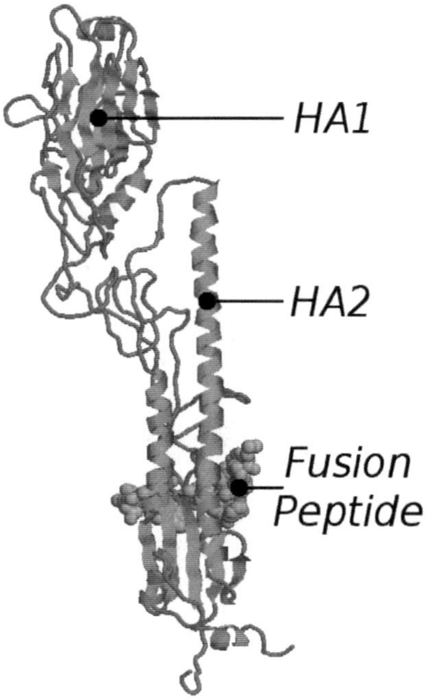

Figure 1. A scheme of the monomer of hemagglutinin in a native state. The monomer consists of two domains, HA1 and HA2. The secondary structure of HA1 domain is mainly formed by β-sheets. The secondary structure of HA2 domain is mainly formed by α-helices; three successive regions in HA2 polypeptide chain are able to form stable secondary structure as α-helices, yet only the outermost parts form α-helices in a native state. The smaller of these helixes ends with the polypeptide region formed mainly of hydrophobic amino acid residues, which allow it to penetrate inside the cell membrane. This region got the name "fusion peptide" [3]. The fusion peptide is marked with the representation of atoms by spheres.

The secondary structure of HA1 domain is mainly formed by β-sheets. Its surface contains the binding site for sialic acids, located on the membrane of target cell [4]. The secondary structure of HA2 domain is mainly formed by α-helices; three successive regions in HA2 polypeptide chain are able to form stable secondary structure as α-helices, yet only the outermost parts form α-helices in a native state [5]. The middle region in a native state fails to form this structure and is separated by HA1 subunits from water molecules, which are able to induce the formation of helices in this region. Thus, the structure of this molecule possesses an additional Gibbs free energy, which is likely to provide for the intramolecular rearrangement [6]. The contact of HA domain with water molecules induces formation of a long α-helix of all the three regions; this α-helix is perpendicular to the virus envelope and protrude from the previous protein boundary towards the possible contact with the cell membrane. The protruding part of this helix ends with the polypeptide region formed mainly of hydrophobic amino acid residues, which allow it to penetrate inside the cell membrane. This region got the name "fusion peptide".

When pH is decreased or temperature is elevated, hemagglutinin changes its conformation. It is assumed that this conformational changes proceeds as follows [7] [8]. At neutral pH, the attractive forces of HA1 subunits in the hemagglutinin trimer compensate the repulsive forces and stabilize the trimer. A decrease in pH leads to an increase in the protonation of subunits, positive shift of their charges, and increase in the repulsive forces. [6] In this process, the subunits of trimer become more distant, thereby forming a cavity sufficient for water molecules to penetrate and hydrate the above mentioned regions in the polypeptide chains of three HA2 subunits, causing an irreversible formation of three α-helices coiled in parallel to the trimer axis [8]. If the conformational changes take place

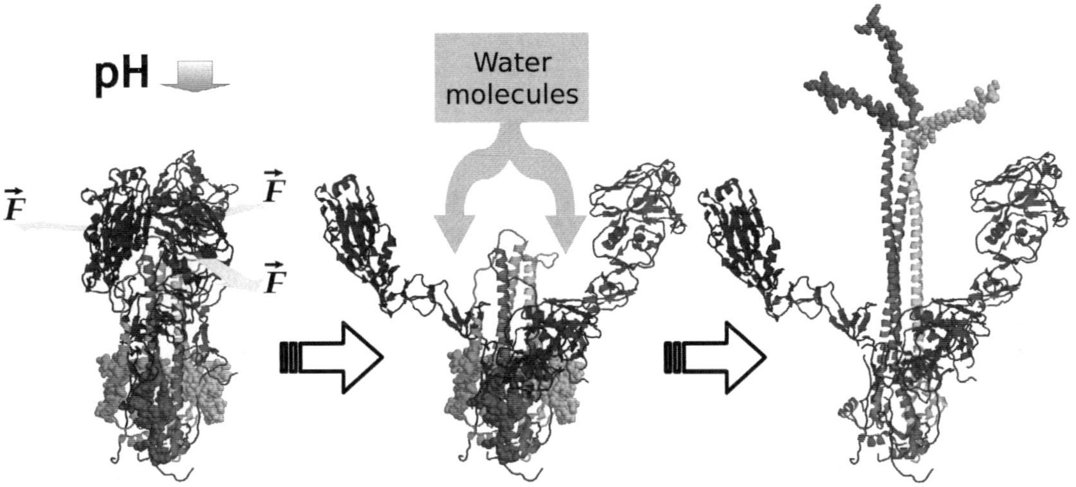

Figure 2. A scheme of conformational changes of the hemagglutinin trimer. A decrease in pH leads to an increase in the protonation of subunits, positive shift of their charges, and increase in the repulsive forces [6]. In this process, the subunits of trimer become more distant, thereby forming a cavity sufficient for water molecules to penetrate and hydrate the above mentioned regions in the polypeptide chains of three HA2 subunits, causing an irreversible formation of three α-helices coiled in parallel to the trimer axis [8]. Fusion peptides are marked with the representation of atoms by spheres.

inside the endosome in the glycoprotein bound in a receptor manner to the cell membrane, these helices penetrate with their fusion peptides into the cell membrane. It is assumed that the state of the trimer having concurrently penetrated into the cell membrane and virus envelope is unstable and induces the changes leading eventually to the fusion of virus envelope and cell endosome membrane followed by formation of the pore connecting the virion with the cell cytoplasm [9]. Thus, it is likely that the most critical and the only reversible stage in the overall process of pH-dependent conformational changes of hemagglutinin molecules in the influenza virus envelope is the initiation, when the possibility is created for water molecules to penetrate into the cavity of hemagglutinin trimer between its subunits.

In this work, we focused on description and study of this stage, which is the consequence of disbalance between the forces stabilizing the structure of hemagglutinin trimer.

As the disulfide bond in the part of the molecule closest to the virus envelope is the only covalent bond between HA1 and HA2 domains and it does not prevent the deviation of HA1 subunit from the trimer axis [3], then the dense folding of hemagglutinin molecule in the trimer, existing under normal conditions, must be determined by van der Waals and electrostatic forces and the interactions of amino acid residues with the solvent. HA1 and HA2 subunits contain a large number of amino acid residues accessible to water and capable of being charged. Therefore, the changes in electrostatic forces caused by a decrease in pH must considerably contribute to the change in the force balance within the trimer [6]. As the hydration of previously inaccessible protein parts completes the initiation stage, we can consider that the area of water contact at this stage is constant, allowing us to neglect the contribution of solvation to the change in the force balance within the trimer. We can also neglect the contribution of van der Waals interaction, as no considerable molecular movements take place at the initiation stage. Thus, we can infer that the electrostatic force is the main contributor to the disbalance that initiates hemagglutinin conformational changes. The value of this force is determined by the equilibrium constants of water dissociation, which, in turn, depends on the reaction temperature.

The main goal of our study was to construct a model making it possible to assess the critical pH value characterizing the fusogenic activity of influenza virus hemagglutinin from the data on hemagglutinin structure and experimental verification of this model. Under this model, we assume that when the electrostatic force between interacting hemagglutinin molecules in the virus envelop exceeds a certain value, hemagglutinin HA1 subunits are arranged so that they form a cavity sufficient for penetration of water molecules. This event leads to an irreversible hydration of the inner fragments of hemagglutinin molecule in the trimer and to completion of conformational changes.

Materials and methods

Computer modeling of the process. A popular model describing the protein behavior in aqueous solutions forms the background for the modeling procedure and calculation of the protein interactions, accomplished as original software; in this model, it is assumed that the molecules in question carry electrostatic charges acquired due to redox reactions [10] [11] [12]. In the developed algorithm, the charged amino acid residues were modeled as point charges placed at the center of the most electronegative atom involved in water dissociation. The geometry of electrostatic field strength for the protein molecule is determined by the geometry of its surface and pH of solution. The geometry of molecule surface is the interface between two dielectrics, protein globule and water, which influences the electrostatic field due to polarization effects [12]. These effects can considerably contribute to the interactions of globular proteins displaying a domain structure. The pH of solution encompassing protein molecule also influences the geometry of electrostatic field, as the values for isoelectric points for various amino acid residues differ considerably. The electrostatic charge of amino acid residues within the protein molecule was calculated assuming that an atom could be either dissociated in the presence of water or not. Here we used the values of isoelectric points of amino acid residues [13], also taking into account the dipole amino acid pairs formed during protein folding when the structure is stabilized by the bonds between oppositely charged amino acid residues. To find the critical pH values, we used the introduced concept of the threshold force as a minimal electrostatic repulsive force of hemagglutinin subunits necessary for formation of the cavity with size sufficient for water molecule penetration into the hemagglutinin trimer on the influenza virus envelope.

Study of hemolysis kinetics. The hemolysis of erythrocytes in the presence of influenza virus particles [14] was used for the experimental study of the kinetics of pH-dependent hemagglutinin conformational changes. In this approach, the endosome membrane is simulated by the erythrocyte membrane, covered with sialic acids, which provide the absorption of influenza virus virions.

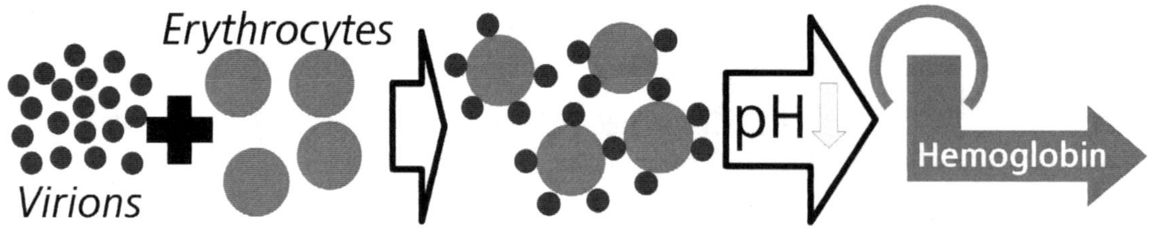

Figure 3. A scheme of the hemolysis of erythrocytes in the presence of influenza virus particles. At the first stage of this reaction, erythrocytes are hemagglutinated due to the sorbed virus particles. At the second stage, pH of the solution with agglutinated erythrocytes is artificially decreased, which changes the conformation of hemagglutinin molecules on the virus envelope and leads to lysis of erythrocytes.

At the first stage of this reaction, erythrocytes are hemagglutinated due to the sorbed virus particles [15] [16]. At the second stage, pH of the solution with agglutinated erythrocytes is artificially decreased, which changes the conformation of hemagglutinin molecules on the virus envelope and leads to lysis of erythrocytes. The degree of lysis is determined spectrophotometrically according to the measured concentration of hemoglobin released into solution depending on the reaction duration. It is very important to control the temperature of hemagglutination and hemolysis of erythrocytes. When calculating the rate of erythrocyte hemolysis, we assumed that (1) all virus particles were absorbed on erythrocytes; (2) the conformational changes of influenza virus hemagglutinin were described by the equation for a simple irreversible chemical reaction where the rate constant is a single-valued function of temperature and pH; and (3) completion of the conformational changes obligatory causes erythrocyte hemolysis. The concentrations of erythrocytes and virus particles were selected so that each erythrocyte was bound to no more than one virion, as this provided that the amount of hemoglobin released from erythrocytes was proportional to the amount of hemagglutinin trimers involved in the reaction. The kinetic constant of hemolysis was calculated according to the following equation:

$$C = C_0 \cdot \left(1 - e^{-kt}\right) \qquad (1)$$

where ***k***, reaction kinetic constant; ***t***, reaction time; $\boldsymbol{C_0}$, concentration of all virus receptors on erythrocyte membranes; and ***C***, concentration of the reacted virus receptors on erythrocyte membranes.

The following influenza virus strains were used in the work: A/Aichi/2/86 (H3N2), obtained from the collection of microorganisms with the FSRI SRC VB VECTOR, and A/NIB/23/89 M and A/NIB/23/89 MA, kindly provided by Dr. A.S. Gambaryan.

Results and discussion

Computer modeling of the interaction between hemagglutinin molecules within the trimer makes it possible to calculate the dependence of the force and the moments of force acting on HA1 subunit from the trimer on the pH value. We calculated the force projection onto the axis directed perpendicular to the symmetry axis from the trimer center to the center of mass of HA1 subunit. The values of threshold force for proteins of the influenza virus strains belonging to the same subtypes were considered equal. The calculations were performed using the data on amino acid sequences of the hemagglutinin molecules of influenza virus strains A/NIB/23/89 M (H1N1), A/NIB/23/89 MA (H1N1), A/FM/1/47 (H1N1), A/Aichi/2/86 (H3N2), and A/HK/1/86 (H3N2) and the spatial model of hemagglutinin molecule extracted from the SWISS-PROT database. Figure 4 shows the calculations of the forces acting on HA1 subunit.

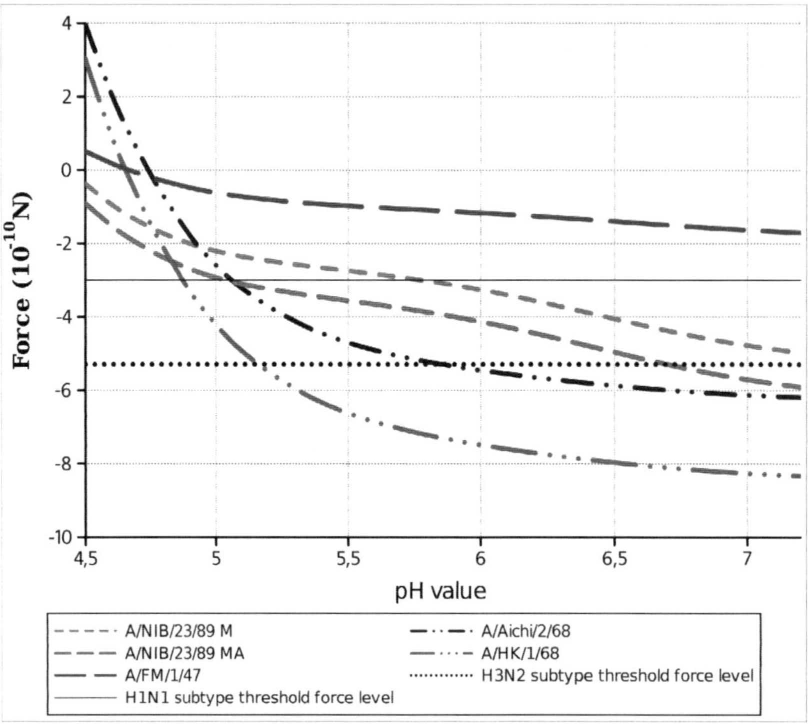

Figure 4. Calculated dependence of projections of the forces acting on HA1 subunit from the trimer on pH of the medium. The force is projected onto the axis directed perpendicular to the symmetry axis from the trimer center to the center of mass of HA1 subunit.

Figure 5 shows the experimental measurements of the dependence of kinetic constants of erythrocyte hemolysis on pH of the medium for one of the studied influenza virus strains. It is evident that the rate of erythrocyte hemolysis increases with temperature.

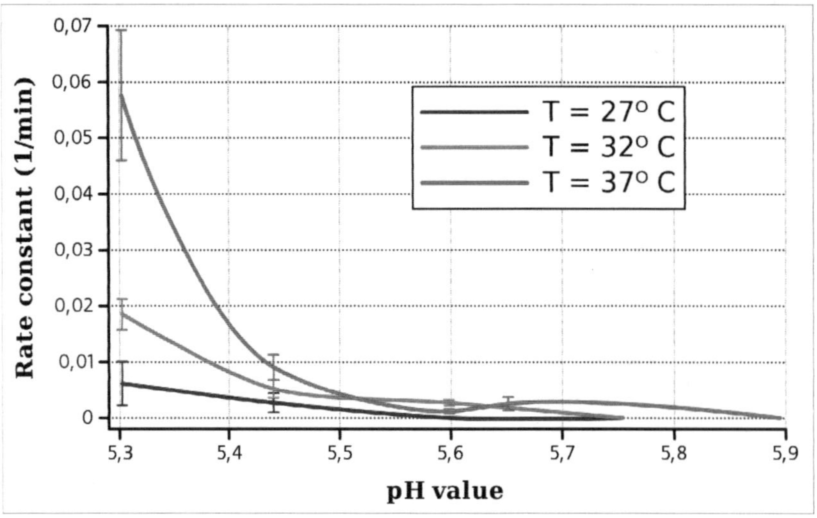

Figure 5. Kinetic constants of erythrocyte hemolysis versus pH of the medium at +27, +32, and +37°C for influenza virus strain A/NIB/23/89 MA.

The theoretically calculated and experimentally measured critical pH values determining the beginning of conformational changes in hemagglutinin molecule for various influenza virus strains are listed in Table 1. The experimental results for pH-dependent erythrocyte hemolysis in the presence of influenza virus strains A/FM/1/47 (H1N1) and A/HK/1/86 (H3N2) were taken from [17] and [18]. The critical pH values for these strains were determined by extrapolating the curves determining the dependence of hemolysis degree on pH value to their intersection with the abscissa. The theoretical calculations for these strains were based on the hemagglutinin amino acids sequences kindly provided by Dr. E. Brown.

Table 1. Theoretically calculated and experimentally measured critical pH values for various influenza virus strains (asterisks indicate the values obtained using the experimental data of [17] and [18]).

Strain	Critical pH values	
	Experimental	Theoretical
A/NIB/23/89 M	5,7 ± 0,1	5,7
A/NIB/23/89 MA	5,4 ± 0,1	5,1
A/FM/1/47	6,0 *	> 7,0
A/Aichi/2/68	5,9 ± 0,1	5,9
A/HK/1/68	5,2 *	5,2

All the experimental dependences of kinetic constants for erythrocyte hemolysis decrease monotonically with the increase in pH and are well described with the equation of irreversible chemical reaction. The obtained dependences do not contradict the hypothesis underlying the model in question, namely, the assumption that the conformational changes are initiated by an increase in the electrostatic repulsive forces with the decrease in pH. The calculation of critical pH values is based on the introduced concept of the threshold force, which is a minimal electrostatic repulsive force between hemagglutinin subunits necessary for formation of the cavity with a size sufficient for penetration of water molecules. The penetration of water molecules into the cavity within the trimer leads to hydration of inner regions in hemagglutinin molecules, formation of helical region in HA2 subunits, and irreversible conformational rearrangement of all the three molecules. The critical pH values obtained by computer modeling and experimental measurements fit sufficiently well (see Table 1), thereby convincingly confirming the applicability of the developed model to calculation of protein interactions.

Thus, the main result of this work is the model allowing the function of a virus to be predicted based on the structure of its molecule, i.e., it allows the critical pH value characterizing the fusogenic activity of influenza virus necessary for delivering the virus genome into the host cytoplasm and infecting the host cell to be assessed from the amino acid sequence of its hemagglutinin. Further development of this model will make it possible to assess biological properties of influenza virus strains, in particular, their infectivity, based on the sequences of virus genome

Acknowledgments

We thank Dr. A. S. Gambaryan for providing with strains A/NIB/23/89 M and A/NIB/23/89 MA and their hemagglutinin amino acids sequences. We thank Dr. E. G. Brown for providing with hemagglutinin amino acids sequences for the strains A/FM/1/47 A/HK/1/68. This research was partly supported by International Scientific and Technological Center grant #2464 and Biotechnology engagement program grant #54.

References

[1] Hernandez L. D., Hoffman L. R., Wolfsberg T. G., White J. M. 1996 Virus-cell and cell-cell

fusion. *Annu Rev Cell Dev Biol.* **12** 627–661

[2] Hamm-Alvarez S. F. 1999 Targeting endocytosis and motor protein to enhance DNA persistence. *Pharmaceutical Science & Technology Today* **2** 190-196

[3] Sauter N. K., Hanson J. E., Glick G. D., Brown J. H., Crowther R. L, Park J., Skehel J. J., Wiley D. C. 1992 Binding of influenza virus hemagglutinin to analogs of its cell-surface receptor, sialic acid: analysis by proton nuclear magnetic resonance spectroscopy and X-ray crystallography. *Biochemistry* **31** 9609-21

[4] Matrosovich M., Tuzikov A., Bovin N., Gambaryan A., Klimov A., Castrucci M. R., DonatellI I., and Kawaoka Y. 2000 Early Alterations of the Receptor-Binding Properties of H1, H2, and H3 Avian Influenza Virus Hemagglutinins after Their Introduction into Mammals. *Journal of Virology* **74** 8502–8512

[5] Shental-Bechor D., Danieli T., Henis Y. I., Ben-Tal N. 2002 Long-range effects on the binding of the influenza HA to receptors are mediated by changes in the stability of a metastable HA conformation. *Biochimica et biophysica acta. Biomembranes* **1565** 81-90

[6] Huang Q.,Opitz R.,Knapp E.-W.,and Herrmann A. 2002 Protonation and Stability of the Globular Domain of Influenza Virus Hemagglutinin. *Biophysical Journal* **82** 1050–1058

[7] Madhusoodanan M. and Lazaridis T. 2003 Investigation of Pathways for the Low-pH Conformational Transition in Influenza Hemagglutinin. *Biophysical Journal* **84** 1926-1939

[8] Huang Q., Sivaramakrishna R. P., Ludwig K., Korte T., Böttcher C., Herrmann A. 2003 Early steps of the conformational changes of influenza virus hemagglutinin to a fusion active state. Stability and energetic of the hemagglutinin. *Biochimica et biophysica acta. Biomembranes* **1614** 3-13

[9] Kozlov M. M. and Chernomordik L. V. 1998 A Mechanism of Protein-Mediated Fusion: Coupling between Refolding of the Influenza Hemagglutinin and Lipid Rearrangements. *Biophysical Journal* **75** 1384 –1396

[10] Volkenshtein M. V. 1988 *Biofizika. (Biophysics.)* (in Russian) (Moscow: Nauka)

[11] Volkenshtein M. V. 1975 *Molekularnaya Biofizika. (Molecular Biophysics.)* (in Russian) (Moscow: Nauka)

[12] Finkelshtein A. V. Ptitsin O. B. 2002 *Fizika Belka. (Physics of Protein.) (in Russian)* (Moscow: Moscow State University Press)

[13] Doson R., Elliot D., Elliot U., and Johns C. 1991 *A Handbook for Biochemists.* Russian translation from Drutsa V. L. and Korolyova O. N. (Moscow: Mir)

[14] Meyhl B. V. 1988 *Virusologia. Metody. (Virology. Methods.)* (in Russian) (Moscow: Mir)

[15] McClelland, L. and Hare R. 1941 The adsorption of influenza virus by red cells and a new in vitro method of measuring antibodies for influenza virus in the embryonated egg. *Can J Public Health* **32** 530-538

[16] Hirst G. K. 1941 The agglutination of red cells by allantoic fluid of chick embryos infected with Influenza virus. *Science* **94** 22-23

[17] Brown E. G., Liu H., Chang Kit L., Baird S. and Nesrallah M. 2001 Pattern of mutation in the genome of influenza A virus on adaptation to increased virulence in the mouse lung: Identification of functional themes. *Proceedings of the National Academy of Sciences of the USA* **98** 6883– 6888

[18] Smeenk C. A. ,Wright K. E. ,Burns B. F. ,Thaker A. J. ,Brown E. G. 1996 Mutations in the hemagglutinin and matrix genes of a virulent influenza virus variant, A/FM/1/47-MA, control different stages in pathogenesis. *Virus Research* **44** 79-95

Optical Kerr Effect Measurements on Supercooled Water: the experimental perspective

P. Bartolini[1], A. Taschin[1], R. Eramo[1,2], R. Righini[1,3] and R. Torre[1,4]

[1]LENS, Univ. di Firenze, Via, N. Carrara 1, I-50019 Sesto Fiorentino, Firenze, Italy.
[2]CNR, Piaz. A. Moro 2, I-00185, Roma,Italy.
[3]Dip. di Chimica, Univ. di Firenze, Via della Lastruccia 3, I-50019 Sesto Fiorentino, Firenze,Italy.
[4]Dip. di Fisica, Univ. di Firenze, Via Sansone 1, I-50019 Sesto Fiorentino,Firenze, Italy.

E-mail: torre@lens.unifi.it

Abstract.

The time-resolved spectroscopic investigations are particularly suitable for the measurement of relaxation patterns of complex liquids where several dynamic phenomena, taking place on different time scales, are present. In particular, the optical Kerr effect performed with ultra-short laser pulses and heterodyne detection enables a unique investigation of the dynamic processes taking place in molecular liquids of different natures.

We present here an improved experimental set-up that enables to measure the fast vibrational dynamics and the slow structural relaxation in a single spectrum characterized by low noise and wide dynamic range. We report one measurement of water dynamics in its supercooled phase.

1. Introduction

Time-resolved non-linear spectroscopy, during the last years, showed to be a unique tool for the investigation of dynamic phenomena in complex liquids [1]. In particular, the time-resolved optical Kerr effect (OKE) investigations [2] enable the measurement, in a single high quality data, of the whole dynamics inclusive of the fast vibrational processes and of the slow structural relaxation phenomena. With the implementation of the heterodyne detection (HD-OKE) and thanks to the new reliable laser sources, HD-OKE is becoming a leading technique in the studies of liquids and complex fluids [3].

Several time-resolved HD-OKE studies have been performed on liquid of different natures [3], as for example simple molecular liquids [4] and glass-formers [5, 7, 8]. Also the water dynamics have been investigated by HD-OKE experiments [9, 10, 11, 12, 13] but only one work has been extended to the water supercooled phase [14]. All the HD-OKE investigations of liquid water have shown the presence of a fast vibrational dynamics followed by a slower monotonic relaxation. The initial oscillatory component of the OKE signal provides information about intermolecular hydrogen bond dynamics, whose microscopic origin has still many unclear aspects. The slower decay has been addressed to structural/diffusive relaxation phenomena [14, 15], whose features are particularly evident in the supercooled phase. No detailed HD-OKE investigations of the vibrational dynamics in supercooled water has been carried out, due to the intrinsic experimental difficulties. Indeed the OKE experiment on supercooled water presents two main difficulties:

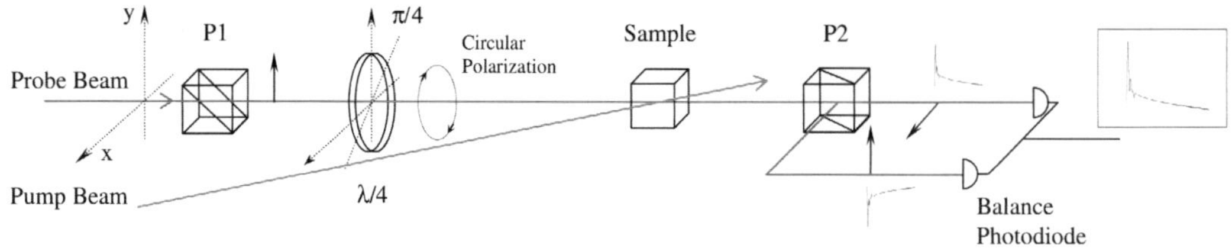

Figure 1. Scheme of the optical set-up used in the HD-OKE experiment. P1 and P2 are the polarizers and $\lambda/4$ is a quarter-wave-plate.

the very low level of OKE signal produced by liquid water and the intrinsic metastability of the supercooled phase.

Using an improved HD-OKE experimental set-up, we have been able to measure the fast vibrational dynamics and the slow structural relaxation in a single spectrum characterized by low noise and wide dynamic range. Moreover, we succeeded in measuring the dynamics of supercooled water over a large time interval with a high time resolution. We can thus provide an unambiguous measurement of the entire time-dependent correlation function of the water in the supercooled phase.

2. Experimental Set-Up

The time-resolved OKE spectroscopy is a pump-probe technique based on the impulsive generation of a transient birefringence by a laser pulse in an optically isotropic sample. When the pump laser pulse passes through the sample it induces a modification of the optical properties (i.e. a transient optical anisotropy or birefringence). This is a transient effect that relaxes back to the equilibrium state through a variety of processes. A second laser pulse, the probe, is sent through the sample. The transient birefringence, induced by the pump in the sample, modifies the probe polarization state. Since the probe pulse arrives with a controlled time delay, it monitors and measures the transient optical excitation and hence the relaxation processes. Typically, in the time-resolved OKE both the pulses, pump and probe, are polarized and non-resonant with any electronic state.

There are different optical set-up that can be used to perform an OKE experiment [3]. Here we describe our latest experimental set-up based on an improved optical scheme [16] and on an innovative acquisition system [17]. This OKE experiment has shown to be particularly successful to detect low intensity signals in the supercooled water. In the figure 1 we report the optical scheme used. In this scheme, the fast axis of $\lambda/4$ wave-plate is rotated of $\pi/4$ in order to produce a circularly polarized probe field. The two signals, corresponding to the horizontal and vertical linear polarizations generated by polarizer P2, are sent to the balanced photodiode detector. The OKE signal measured in such a way is automatically heterodyned and free from eventual dichroic contributions [16]. A further improvement of the OKE signal is obtained subtracting the two HD-OKE measurements, corresponding at left and right circular polarizations of the probe field (i.e. $\pm\pi/4$ positions of the wave-plate axis). This procedure removes from the signal the spurious dichroic contributions coming from possible misalignment of the wave-plate (i.e. angle of rotation different from $\pi/4$). Here we are using a real-time acquisition of the experimental signal during the rapid scan of the optical delay line [17]. This acquisition scheme enables data acquisition with high absolute time resolution, $\approx 1\ fs$, high scanning velocity, $\leq 2.5\ [cm \cdot s^{-1}]$, and long time delay, few ns. In figure 2 we report the complete experimental scheme. The laser is a high repetition-rate Kerr-lens mode-locked Ti:sapphire laser (FemtoRose 10MDC, by R&D Ultrafast Laser) pumped by the second harmonic of a laser-diode-pumped Nd:YVO_4 laser

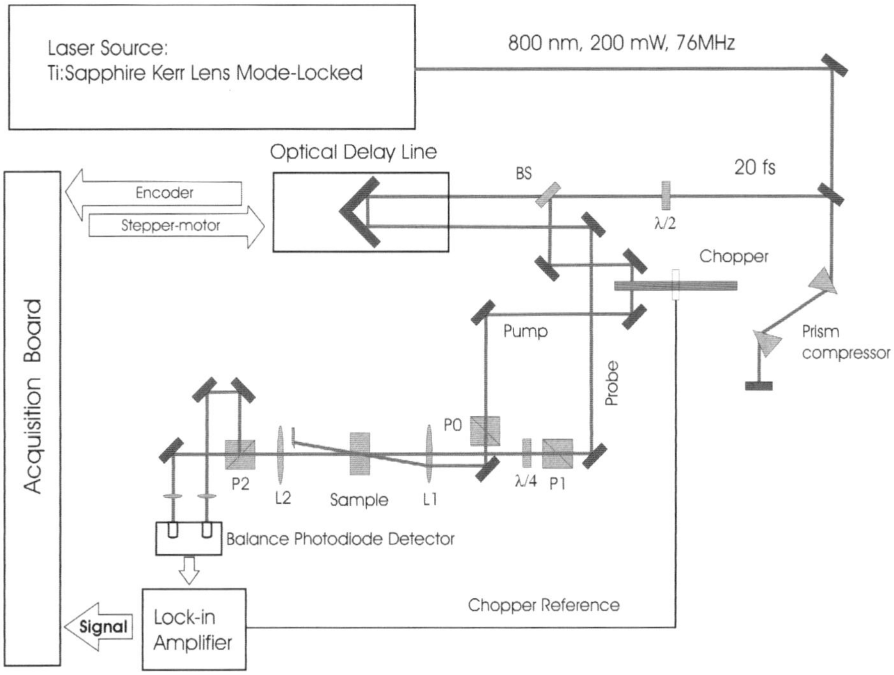

Figure 2. The complete experimental set-up for HD-OKE spectroscopy, inclusive of the laser source and the real-time data acquisition scheme.

(Millennia V, by Spectra-Physics) with a pulse width of 20 fs [17].

3. HD-OKE signal on supercooled water

The signal beam measured in an HD-OKE experiment is [3, 5, 7]:

$$S_{hd}(\tau) \propto \int dt \; |E_{pr}(t-\tau)|^2 \int dt' \mathcal{R}_{oke}(t-t')|E_{ex}(t')|^2 \tag{1}$$

being E_{pr} and E_{ex} the probing and exciting laser pulses, respectively; $R_{oke}(t)$ is the material response function. Since we are performing a non-resonant OKE experiment, the Born-Oppenheimer approximation applies and the response function becomes [3]:

$$\mathcal{R}_{oke}(q,t) \propto -\frac{\theta(t)}{k_B T}\frac{\partial}{\partial t}\langle \chi_{ij}(t)\; \chi_{ij}(0)\rangle \tag{2}$$

Where χ_{ij} is linear susceptibility, $\langle\cdot\rangle$ is the ensemble average, T is the temperature, k_B is the Boltzmann constant and $\theta(t)$ the Heaviside step function. $\mathcal{R}(t)$ defines all the material dynamic information.

In figure 3 we report, as example, a HD-OKE signal obtained in a supercooled water sample at a temperature of $T = -24\ ^oC$. This HD-OKE experiment enables the measurement of the fast vibrational dynamics and the slow structural relaxation in a single data, from about 20 fs to 35 ps. This data confirms the presence of a non-exponential monotonic decay, from about 1 ps to 35 ps, that has to be addressed to the slow structural/diffusive relaxation [14]. In the fast time scale, up to about 1 ps,the HD-OKE signal displays a complex oscillatory pattern. This signal features are due to the inter-molecular vibrational dynamics of the hydrogen bound structure.

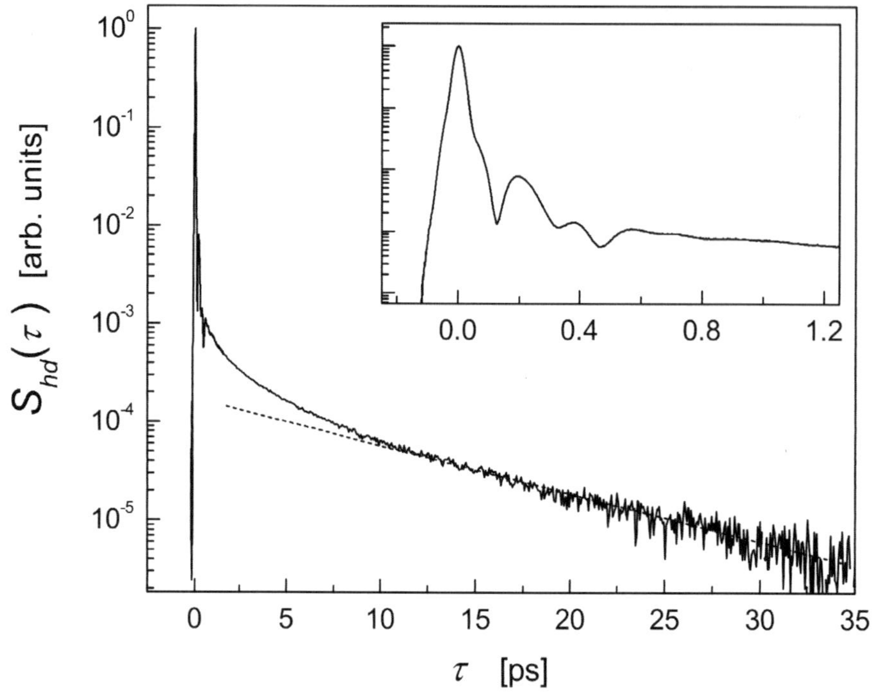

Figure 3. HD-OKE signal on water at $T = -24\ ^oC$ obtained with the HD-OKE experimental set-up reported in figure 2. The inset shows some details of the vibrational fast response. The dashed line is a simple exponential decay, superimposed as response.

The research has been performed at LENS, supported by the EC grant N. RII3-CT-2003-506350.

References

[1] Torre R 2008 *Time-resolved spectroscopy of complex liquids* (Springer, New York)
[2] Righini R 1993 *Science* **262** 1386
[3] Bartolini P, Taschin A, Eramo R and Torre R 2008 *Time-resolved spectroscopy of complex liquids* ed R Torre (Springer, New York) p 73
[4] Ricci M, Wiebel S, Bartolini P, Taschin A, and Torre, R 2004 *Philos.Mag.* **84** 1491
[5] Torre R, Bartolini P, Ricci M, Pick R M 2000 *EuroPhys.Lett.* **52** 324
[6] Ricci M, Bartolini P and Torre R 2002 *Philos.Mag.B* **82** 541
[7] Prevosto D, Bartolini P, Torre R, Ricci M, Taschin A, Capaccioli S, Lucchesi M and Rolla P 2002 *Phys.Rev.E* **66** 011502
[8] Pratesi G, Bartolini P, Senatra D, Ricci M, Righini R, Barocchi F and Torre R 2003 *Phys.Rev.E* **67** 021505
[9] Palese S, Schilling L, Miller R J D, Staver P R and Lotshaw W T 1994 *J. Phys. Chem.* **98** 6308
[10] Castner E W, Chang Y J, Chu Y C and Walrafen G E 1995 *J. Chem. Phys.* **102** 653
[11] Palese S, Mukamel S, Miller R J D, Lotshaw W T 1994 *J. Phys. Chem.* **100** 10380
[12] Winkler K, Lindner J, Buoersing H and Voehringer P 2000 *J.Chem.Phys.* **113** 4674
[13] Winkler K, Lindner J and Voehringer P 2002 *Phys. Chem. Chem. Phys.* **4** 2144
[14] Torre R, Bartolini P and Righini R 2004 *Nature* **428** 296
[15] Bellissent-Funel M C, Longeville S, Zanotti J M and Chen S-H 2000 *Phys. Rev. Lett.* **85** 3644
[16] Giraud G, Gordon C M, Dunkin I R and Wynne K 2003 *J. Chem. Phys.* **119** 464
[17] Bartolini P, Taschin A, Eramo R and Torre R 2007 *Phil. Mag.* **87** 731

Journal of Physics: Conference Series **177** (2009) 012010 doi:10.1088/1742-6596/177/1/012010

Structural studies of water in a confined hydrophobic environment

JC Dore[1], M-C Bellissent-Funel[2], A Burian[3], HP Castricum[4], J Jelassi[5], K Kaneko[6], T Ohba[6], H Tanaka[7] and JBW Webber[1,8]

1. School of Physical Sciences, University of Kent, Canterbury CT2 7NH
2. Laboratoire Léon Brillouin, CEA, Saclay, France
3. A.Chelkowski Institute of Physics, University of Silesia, ul. Uniwersytecka 4 40-007 Katowice, Poland
4. Van't Hoff Institute for Molecular Sciences, Universiteit van Amsterdam Nieuw Achtergracht 166, 1018 WV Amsterdam, The Netherlands
5. Dept de Physique, Université de Monastir, Tunisia
6. Chemistry Dept, Chiba University, Inage, Chiba 263-8522, Japan
7. Department of Chemical Engineering, Kyoto University, Kyoto 615-8510 Japan
8. Institute of Petroleum Engineering, Heriot Watt University, Edinburgh EH14 4AS.

E-mail:J.C.Dore@kent.ac.uk

1. Introduction

Water has been studied under a wide range of conditions and much current interest centres on water in the confined geometry of a mesoscopic solid. Most investigations have used mesoporous silicas in the form of a sol-gel or templated materials such as MCM41 and SBA-15. In these cases, the interface is hydrophilic due to the hydrogen-bonding of water molecules across the interface to silanol groups at the silica surface. The properties determined from neutron diffraction and NMR studies have been reviewed in 2004 by Webber and Dore [1]. However, it is also of interest to investigate the properties of water at a hydrophobic interface but these measurements are more difficult due to several technical problems linked to the adsorption of the water into the pore volume.

2. Experimental measurements

There are two different approaches to this question. The first involves the modification of the silica interface by grafting a layer onto the internal surface by chemical means. This technique has been used by Castricum and colleagues [2] for a sol-gel silica of ~55Å pore size in which the internal surface is coated with a 'methylated' layer. The second uses the hydrophobic features of a graphene sheet, as water may be readily adsorbed into carbon nanotubes. A convenient material for this work is based on a carbon nanohorn sample which has been used extensively by Kaneko and colleagues [3]. In both cases it must be assumed that there are some active sites in the pore volume that can adsorb small water droplets that subsequently grow to occupy the whole of the

Journal of Physics: Conference Series **177** (2009) 012010 doi:10.1088/1742-6596/177/1/012010

pore volume.
Neutron diffraction measurements for water/ice in the sol-gel samples were made as a function of temperature on the 7C2 diffractometer at the Orphee reactor [LLB, Saclay] and for water/ice in the carbon nanohorns on the GEM instrument at the ISIS Spallation Source [Rutherford Appleton Lab]. The diffraction patterns obtained are illustrated in Figs 1 and 2, respectively. Complementary measurements were made by gas adsorption, DSC, SANS and NMR..

3. Results and Interpretation

The results for the water/ice in the sol-gel silicas showed that there was an initial nucleation as hexagonal ice for 'almost-filled' [f~0.95] pores and a growth of cubic ice at lower temperatures but no hexagonal ice was created for lower filling-factors. There were significant differences between the modified and non-modified samples and these differences became more pronounced for lower water content [f ~0.2]. The water in the carbon nanohorns was found to give a main diffraction peak that was much lower than that observed for bulk water and was not dependent on the temperature. This surprising result indicates a higher degree of hydrogen-bonding and longer-range spatial correlations, suggesting the presence of a more stable H-bonded network. Comparison in real space indicates that the density is much lower [~20%]. It seems that the inability to form hydrogen-bonds across the solid-liquid interface leads to a more structured region for the water in proximity to the interface. Nucleation [<250K] gives hexagonal ice peaks superimposed on a broader distribution characteristic of amorphous ice. A more detailed presentation of the results will be published elsewhere [4,5] and further work, incorporating dynamical studies by QENS and DINS is continuing.

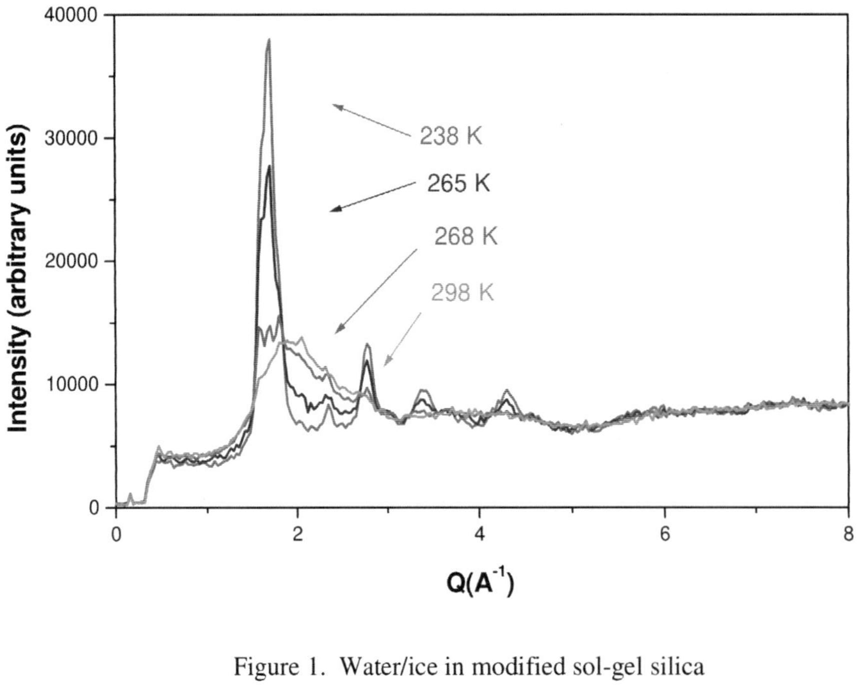

Figure 1. Water/ice in modified sol-gel silica

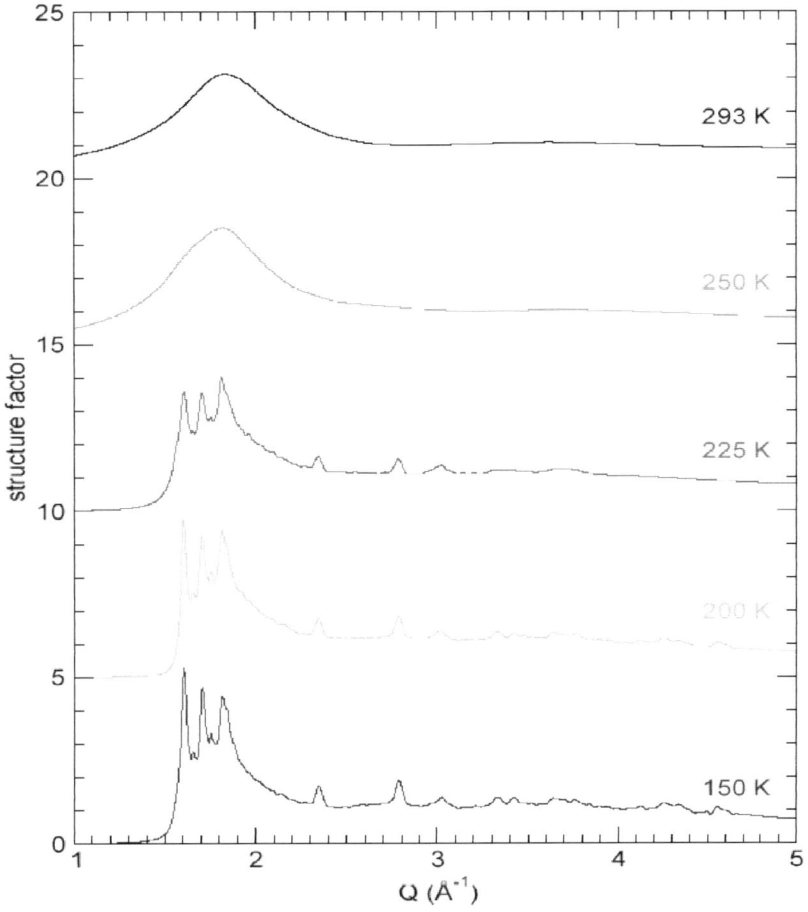

Figure 2. Water/ice in carbon nanohorns

4. Acknowledgements
We wish to acknowledge support from several research-funding bodies that have enabled this work to be carried out in two wide-ranging international collaborations. We also wish to thank Alex Hannon at the ISIS Spallation Neutron Source [Rutherford Appleton Lab] for assistance with the experimental work on the GEM diffractometer and subsequent analysis.

References:
1. J.B.W.Webber and J.C.Dore 2004 *J.Phys:Condens Matt* **16** S5449,
2. H.L.Castricum et al 2006 *Micro- and Mesoporous Materials* **88** 63
3. K.Kaneko, see this conference report and references therein
4. J.Jelassi, H.L.Castricum, M.-C.Bellissent-Funel, J.C.Dore, J.B.W.Webber and R. Sridi-Dorbez, submitted to PCCP
5. J.C.Dore, J.Jelassi, H.L.Castricum, M.-C.Bellissent-Funel, I.Bako and R. Sridi-Dorbez, in preparation

Journal of Physics: Conference Series **177** (2009) 012011 doi:10.1088/1742-6596/177/1/012011

Dynamic transition and glassy behaviour in hydrated proteins

F.Mezei[1,2], M. Russina[3], G. Chen[1], H. Frauenfelder[1], P. W. Fenimore[1], P. Falus[4], B. Farago[4]

[1]*Los Alamos National Laboratory, Los Alamos, NM 87545, USA,*
[2]*Research Institute for Solid State Physics, 1525 Budapest, POB. 49, Hungary*
[3]*Helmholtz-Zentrum-Berlin, Glienicker str 100, 14109 Berlin, Germany*
[4]*Institut-Laue-Langevin, BP 156, 38042 Grenoble Cedex 9, France*

E-mail: mezei@lanl.gov

The temperature dependence of functional kinetics of hydrated proteins has been related to changes in dynamic behaviour on the microscopic level, as reflected by the variation of the mean square displacement of the hydrogen atoms $<u^2>$. The marked kink in the temperature dependence of this quantity around 200 K has been originally observed by Mössbauer spectroscopy and by "elastic window" neutron backscattering spectroscopy [1,2]. This behaviour is referred to as "dynamic transition", although it is clear that it is a gradual change in the temperature dependence as opposed to a sharp feature. Its origin remained controversial over the past decades. We have extended its exploration by a detailed study of the whole quasielastic and inelastic neutron scattering spectra over the broad energy transfer domain of 50 neV to 20 meV. This was achieved by the combination of high precision time-of-flight neutron spectroscopy (TOF) and neutron spin echo (NSE) experiments in both H_2O and D_2O hydrated myoglobin. Around the temperature of the dynamic transition we have observed the onset of a relaxation process spread over several orders of magnitude in time, cf. Fig. 1, which reveals all neutron scattering signature properties of the canonical β relaxation process in the vicinity conventional glass transitions. (All figures 1-3 show new data, to be published in detail elsewhere by the present authors.)

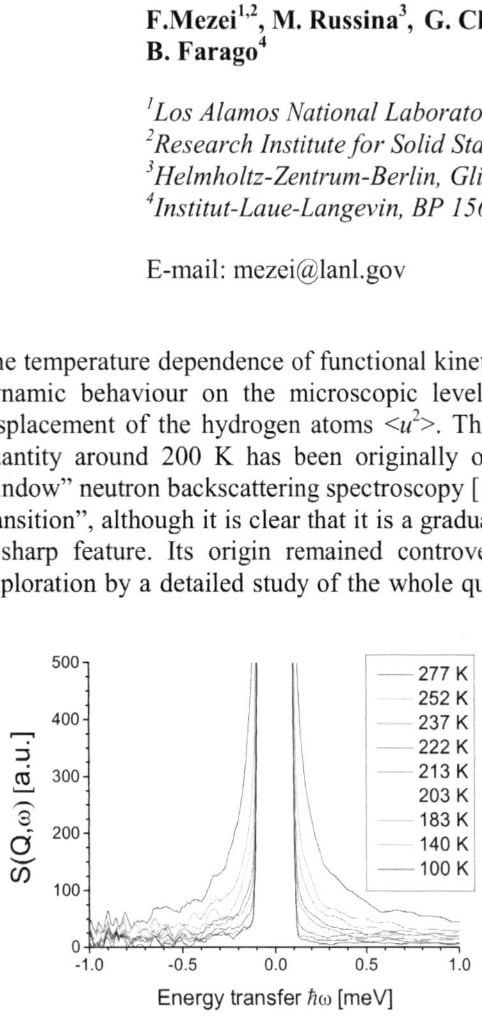

Figure 1. Temperature dependence of the quasi-elastic neutron scattering spectra at momentum transfer Q ≈ 1 $Å^{-1}$ in hydrated myoglobin (0.8 g D_2O for 1 g protein). TOF neutron spectroscopy data obtained by the spectrometer NEAT at HZ Berlin. (The curves show averages over 5 adjacent data points.)

The similarity is reciprocal. We have also found that if analyzed the same way as protein data have been, the quasielastic neutron scattering in glasses also shows archetypical dynamic transition near the glass transition temperature, i.e. a clear kink in the temperature dependence in the mean square displacement $<u^2>$, derived by the standard analysis of the wave-number Q dependence of the Debye-Waller factor observed in the incoherent scattering. It is important to keep in mind that in general the mean square displacement depends on the time t over which the average $<u^2>$ is taken. In our TOF experiments this time was defined either by the energy resolution of the instrument or by integrating the spectra over the energy domain corresponding to the selected averaging time: $-h/t < E < h/t$. The absence of the signature kink of the apparent dynamic transition in Fig. 2 for the shortest averaging time in the canonical glass glycerol indicates that the dynamic process behind the change of $<u^2>$ takes place at times longer than 1 ps.

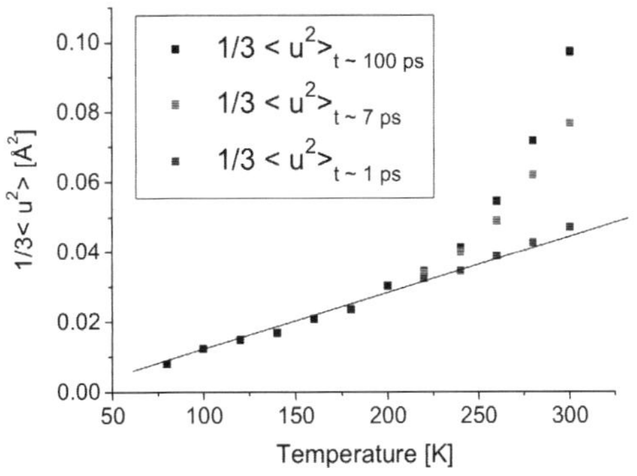

Figure 2. Temperature dependence of the mean square displacement in the archetypical glass glycerol-15 % water, as determined for different averaging times from TOF neutron scattering data similar to the ones shown in Fig. 1. The straight line is guide to the eye

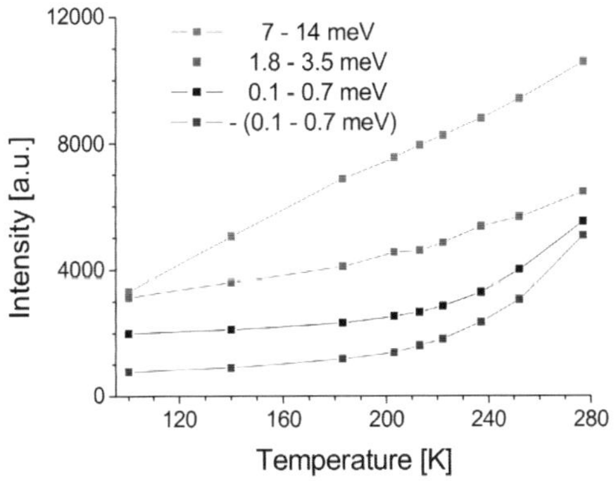

Figure 3. Temperature dependence of the inelastic scattering intensities integrated over various energy transfer domains, for the spectra on hydrated myoglobin partially shown in Fig. 1. The slight downturn above 200 K with increasing temperature in the curve for the 7 – 14 meV bracket is consistent with the drop in the Debye-Waller factor caused by the upturn of the scattering intensities at the lower energy brackets. Only the deviation of the curves from harmonic straight lines is of significance.

For a fully model independent approach to analysing the experimental observations and also getting a more detailed picture, we have examined the temperature dependence of the scattering spectra at different energies, instead of just looking at the mean square displacement produced by all inelastic scattering processes. The results in Fig. 3, as expected, provide an alternative and much more telling signature of the behaviour without any model fitting. The absence of the dynamic transition type kink at the highest energy bracket, which corresponds to phonon type vibrational motion, offers clear evidence that no softening of the elastic constants occurs. This is in contrast to one of the general explanations envisaged for the transition. In contrast, the process appearing at the lower energies is clearly seen to be related to the observed transitional behaviour. Our NSE data show that this process extends over several orders of magnitude of energy/time down to the instrumental resolution of 50 neV and reveal a complex, somewhat temperature dependent line shape. In particular the process stretches a much wider time span than a stretched exponential with a stretching parameter in the common range around 0.5. On the other hand, the spectra are fully compatible with the β fluctuations observed in recent dielectric relaxation experiments [3], which enter – with increasing temperature – the 0.2 ps – 10 ns combined TOF – NSE time window of the present study.

[1] Parak F and Knapp E W (1984) *Proc. Natl. Acad. Sci. USA* **81** 7088
[2] Doster W, Cusack S and Petry W (1989) *Nature* **337** 754
[3] Chen G, Fenimore P W, Frauenfelder H, Mezei F, Swenson J and Young R D 2008 *Phil.Mag.* **88** 33

Journal of Physics: Conference Series **177** (2009) 012012 doi:10.1088/1742-6596/177/1/012012

Relation between frequency and H bond length in heavy water: Towards the understanding of the unusual properties of H bond dynamics in nanoporous media

Stanislas Pommeret[1], Raluca Musat, Jean Philippe Renault, Jean-Claude Leicknam[2] and Savo Bratos[2]

[1] * CEA/Saclay, DSM/DRECAM/SCM URA 331 CNRS, Laboratoire de Radiolyse, F-91191 Gif-sur-Yvette, France [2] Laboratoire de Physique Théorique de la Matière Condensée, Université Pierre et Marie Curie-Paris6, UMR 7600 (CNRS), Paris, France.

The published work on H bond dynamics mainly refers to diluted solutions HDO/D_2O rather than to normal water.[1] The reasons for this choice are both theoretical and experimental. Mechanical isolation of the OH vibrator eliminating the resonant energy transfer makes it a better probe of the local H bond network, while the dilution in heavy water reduces the infrared absorption, which permits the use of thicker experimental cells. The isotopic substitution does not alter crucially the nature of the problem. The water dynamics is still the subject of intense theoretical[2] and experimental[3] studies. The length r of an OH...O group is statistically distributed over a large interval comprised between 2.7 and $3.2\ A$ with a mean value $r_0 = 2.86\ A$. Liquid water may thus be viewed as a mixture of hydrogen bonds of different length. Two important characteristics of hydrogen bonding must be mentioned. (i) The OH stretching vibrations are strongly affected by this interaction. The shorter the length r of the hydrogen bond, the strongest the H bond link and the lower is its frequency ω: the covalent OH bond energy is lent to the *OH..O* bond and reinforces the latter. A number of useful relationships between ω and r were published to express this correlation. The one adopted in our previous work is the relationship due to Mikenda[4]. (ii) Not only the OH vibrations, but also the HDO rotations are influenced noticeably by hydrogen bonding. This is due to steric forces that hinder the HDO rotations. As they are stronger in short than in long hydrogen bonds, rotations are slower in the first case than in the second. This effect was only recently discovered, but its existence is hardly to be contested.[5] In the present contribution, we want to revisit the relationship between the frequency of the OH vibrator and the distance OH..O.

Water molecule has three normal modes $\omega_1, \omega_2, \omega_3$, and the OH stretching frequency of HDO corresponds to ω_3. The quantities examined in what follows are $<\Delta\omega_3(r,t)>$ and

[1] T. Elsaesser, H.J. Bakker, *Ultrafast Hydrogen Bonding Dynamics and Proton Transfer Processes in the Condensed Phase*, Kluwer, Dordrecht, 2002.

[2] M. Diraison, Y. Guissani, J-Cl. Leicknam, and S. Bratos, Femtosecond solvation dynamics of water: solvent response to vibrational excitation of the solute, *Chem. Phys. Lett.* **258**, 348-351 (1996). C. P. Lawrence and J. L. Skinner, Vibrational spectroscopy of HDO in liquid D2O. III. Spectral diffusion, and hydrogen bonding and rotational dynamics, *J. Chem.Phys.* **118**, 264-272 (2003). S. A. Corcelli, C. P. Lawrence, J. B. Asbury, T. Steinel, M. D. Fayer, and J. L. Skinner, Spectral diffusion in a fluctuating charge model of water, *J. Chem. Phys.* **121**, 8897-8900 (2004).

[3] J. Stenger, D. Madsen, P. Hamm, E. T. J. Nibbeling and T. Elsaesser, Ultrafast vibrational dephasing of liquid water, *Phys.Rev. Lett.* **87**, 027401 1-4 (2001). C. J. Fecko, J. D. Eaves, J. J. Loparo, A. Tokmakoff and P. L. Geissler, Ultrafast hydrogen bond dynamics in the infrared spectroscopy of water, *Science* **302**, 1698-1702 (2003). S.Yeremenko, M. S. Pshenichnikov and D. A. Wiersma, Hydrogen-bond dynamics in water explored by heterodyne-detected photon echo, *Chem.Phys.Lett.* **369**, 107-113 (2003).

[4] W. Mikenda, Stretching frequency versus bond distance correlation of O---D(H)...Y (Y = N, O, S, Se, Cl, Br, I) hydrogen bonds in solid hydrates, *J. Mol. Struct.*, **147**, 1-15 (1986).

[5] S. Woutersen, U. Emmerichs, H.J. Bakker, Femtosecond mid-IR pump-probe spectroscopy of liquid water: Evidence for a two-component structure, *Science* **278**, 658-660 (1997). G. Gallot, S. Bratos, S. Pommeret, N. Lascoux, J.-C. Leicknam, M. Kozinski, W. Amir, and G.M. Gale, Coupling Between Molecular Rotations and OH..O Motions in Liquid Water: Theory and Experiment, *J. Chem. Phys.*, **117**, 11301 (2002).

$<\Delta\omega_3(r,t)\ \Delta\omega_3(r,t')>$. In these functions, $\omega_3(r,t)$ indicates the value of the frequency ω_3 at time t given that the OH...O bond length was r at time 0. Moreover, $\Delta\omega_3(r,t) = \omega_3(r,t) - <\omega_3>$. In calculating statistical averages, intramolecular vibrations were treated by quantum mechanics and all remaining variables by classical mechanics. The potential employed was the pair-additive extended-simple-point-charge (SPCE) potential of Berendsen et al.[6] The classical calculations were performed as follows. The simulation box contained 255 D_2O molecules and one HDO molecule with periodic boundary conditions. The long-range electrostatic interactions were taken into account by calculating the Ewald sum. The equations of motion were integrated with the help of leap frog algorithm. The thermodynamic point considered here corresponds to T=300 K and ρ = 1.09 g/cm3. The statistical quantities were extracted from two independent runs of 5 ns after equilibration; the time step of the simulation was 0.5 fs. The configurations of the system were saved every 5 fs and the frequency of the OH vibrator computed. The O...O distance r_{OO} between the oxygen atom of the HOD molecule and that of a D_2O molecule can be defined in different ways. Some authors assimilated it to the O...O distance between the HDO and the closest D_2O. We preferred to attribute it to D_2O molecule for which the H...O distance is the smallest in the same pair of molecules. Then, using quantum mechanical perturbation theory, one obtained analytical expressions of $\omega_3(r,\tau)$ in terms of the derivatives of the intermolecular potential with respect the internal vibrations.[7]

Once the instantaneous frequency $\omega_3(t)$ and distance $r_{OO}(t)$ have been computed for a large number of steps of molecular dynamics, one can examine some properties of H bonds. As noted earlier[8], the variables ω_3 and r_{OO} are correlated. This correlation is strong enough to exhibit a well defined peak in the (ω_3 , r_{OO}) plane. It is then possible to compute the mean distance r_{OO} at a given frequency ω_3, and the mean frequency ω_3 at a given distance r_{OO}. These two functions are plotted in Figure 1A. It is clear that, even if they are different, the difference between them is small. Comparing them with the Mikenda relation[4], one finds a good agreement. Note that the empirical relationship was established for solids.

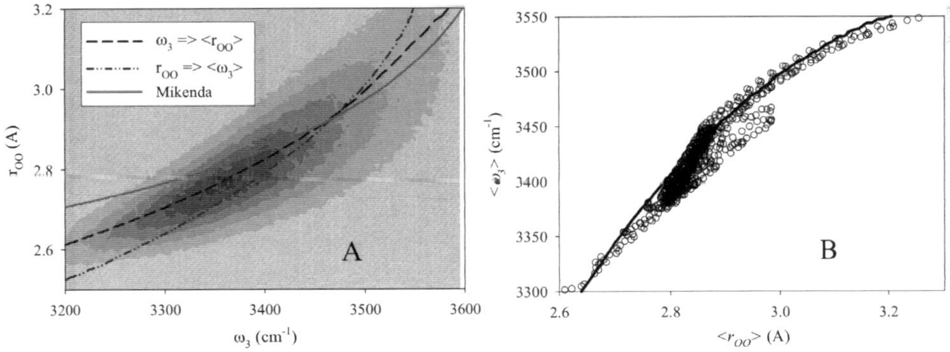

Figure 1: Relationship between the frequency ω_3 of the OH vibrator and the distance r_{OO}. (A) Solid line is the Mikenda relation[4]; the long dash curve gives the mean distance r_{OO} for a given frequency ω_3; and the dash dot dot curve gives the mean frequency ω_3 at a given distance r_{OO}. (B) Parametric plot of $<\omega_3(r,t)>$ as a function of $<r_{OO}(r,t)>$ for times varying between 0 and 3 ps, and r varying between 2.6 and 3.2 A. The solid line is $<\omega_3(r)>$.

[6] H. J. C. Berendsen, J. R. Grigera and T. P. Straatsma, The missing term in effective pair potentials,*J. Phys. Chem.*, **91**, 6269-6271 (1987).

[7] M. Diraison, J-Cl. Leicknam, G. Tarjus, and S. Bratos, Computer simulation study of inelastic neutron scattering from liquid water, Phys. Rev. E 50 2689-2695 (1994).

[8] R. Ray, K. B. Moller and J. T. Hynes, Hydrogen bond dynamics in water and ultrafast infrared spectroscopy, *J. Phys.Chem.A* **106**, 11993-11996 (2002)

Having established a relationship between the frequency and the distance in static conditions, we re-examined it from dynamical point of view. In order to do it, we have computed the mean distance and the mean frequency for different times *t*: we extracted from the MD runs the configurations for which the H bond distance was equal to *r*, and followed the mean frequency $<\omega_3(r,t)>$ and the mean distance $<r_{OO}(r,t)>$ in time. The parametric plot (Figure 1B) shows a strong correlation between these two quantities, whatever the initial value of *r*.
To go further in the comparison between frequency and distance as a function of time, we have developed a new theory that makes the link between the distance and the frequency, assuming that the initial wavepacket is selected by a femtosecond laser pulse. It is inspired by a recent theory of time resolved X-ray diffraction.[9] This theory requires the knowledge of two time dependent functions: $<\omega_3(r,t)>$ that we presented earlier and

$$<\Delta\omega_3(r,t)\ \Delta\omega_3(r,t')>_C = <\omega_3(r,t)\ \omega_3(r,t')> - <\omega_3(r,0)><\omega_3(r,0)> \qquad (2)$$

This function is much alike the standard frequency shift correlation function $<\Delta\omega_3(t)\ \Delta\omega_3,t')>$, but is not identical to it. The frequency shift correlation function is well fitted (Figure 2A) by two exponentials (40 and 625 fs). If t and t' are large with respect to 625 fs, the system has lost the memory of its initial condition and it is easy to show that $<\Delta\omega_3(r,t)\ \Delta\omega_3(r,t')>_C \rightarrow <\Delta\omega_3(t)\ \Delta\omega_3,t')>_C$. In Figure 2B, we illustrated the time dependence of $<\Delta\omega_3(r,t)\ \Delta\omega_3(r,t)>_C$ for r = 2.86 A. For long times, this function reaches the value of $<\Delta\omega_3(0)\ \Delta\omega_3(0)>_C$ = 11780 cm^{-2}, but at short time its value is only 4000 cm^{-2}. This indicates that when pumping at a given frequency, one selects a narrow range of distances of the H bond network. The evolution of the function toward the asymptotic value is characterized, as in the frequency shift correlation function, by a fast and a slow component.

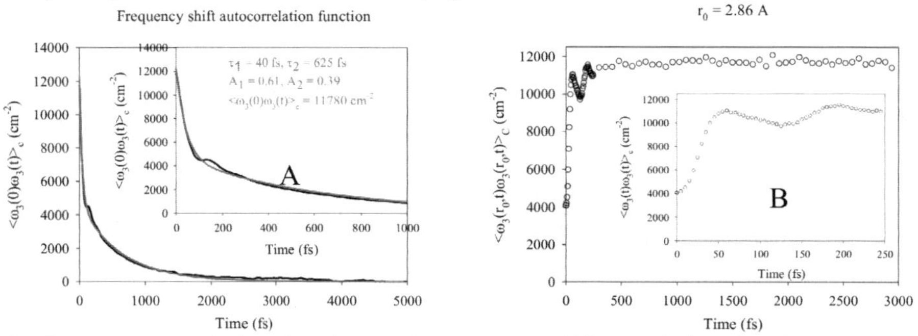

Figure 2: Frequency correlation functions: A) Frequency shift correlation function $<\Delta\omega_3(t)\ \Delta\omega_3(t')>$. B) Function $<\Delta\omega_3(r,t)\ \Delta\omega_3(r,t)>$.

The present study confirms the link between the frequency of an OH vibrator and the H bond network surrounding it. Even so the relation between the distance r_{OO} and ω_3 is statistical in nature; it is sufficiently narrow to affirm that the measurement of the mean frequency ω_3 at a given time is related to the distance r_{OO} at a given time as it was stated earlier.[10]

The present results are currently being expanded to nanoporous media and compared with the recent findings of the CEA team on the unusual properties of the H bond dynamics.

[9] S. Bratos, F. Mirloup, R. Vuilleumier and M. Wulff, Time resolved X-ray diffraction: statistical theory and its application to the photophysics of molecular iodide, *J. Chem. Phys.* **116,** 10615-10625 (2002).
[10] G.M. Gale, G. Gallot, F. Hache, N. Lascoux, S. Bratos, J.-Cl. Leicknam, Femtosecond dynamics of hydrogen bond in liquid water: a real time study, *Phys. Rev. Lett.* **82**, 1068-1071 (1999).

Journal of Physics: Conference Series **177** (2009) 012013 doi:10.1088/1742-6596/177/1/012013

Quantum confinement of hydrogen in ice based clathrates

Margarita Russina[1*], Ewout Kemner[1], Milva Celli[2], Lorenzo Ulivi[2], Ferenc Mezei[3]

[1] Helmholtz Zentrum Berlin, Glienicker Str. 100, 14109 Berlin, Germany
[2] ISC-CNR, via Madonna del piano 10, I-50019 Sesto Fiorentino, Italy
[3] Budapest Neutron Center, SzFKI, Konkoly Thege u 29-33, 1121 Budapest, Hungary

*Margarita.Russina@helmholtz-berlin.de

Clathrate hydrates are inclusion compounds, formed by a network of hydrogen-bonded water molecules that is stabilized by the presence of foreign molecules [1]. The hydrophobic interactions between the host framework and guest molecules push the ice to change its structure and form cages of different form and dimensions to accommodate the guests. The topology of the ice cages depends on the sizes and shapes of the guest molecules, the typical dimensions of the cages range between 4 - 15 Å. The confinement and small size of the cages strongly influence the behavior of the guest molecules and results in properties different from those of the bulk, making clathrates hydrates model system to study the effects of confinement.
The interest in clathrate hydrates was motivated by their potential as cheap and environmentally friendly energy storage carriers. The synthesis of the hydrogen filled ice clathrates has been reported in 2002 by W. Mao [2]. By application of 2000 bar hydrogen pressure the authors were able to form fully hydrogenated clathrates which are stable up to 160 K under atmospheric pressure and can hold hydrogen up to 3.77% of the total mass. It has been found that loading pressure can be decreased (to about 300 bar) and the temperature stability can be improved up to 270 K by introduction of large organic molecules into part of the cages of the H_2-clathrates [3].
To understand the reason of these differences we undertook the detailed study of the guest-host interactions by combined neutron scattering and molecular simulation approach. Guest-host interactions are reflected best in the microscopic dynamics, thus the benefits and complementarities of neutron scattering and molecular dynamics simulations are obvious: both techniques provide information on atomic trajectories allowing for a direct comparison of the theoretical and experimental results. Both methods work with full efficiency on the length scale of few nanometer relevant for hydrogen storage structures and can follow atomic motions on a broad time scale from 10^{-14} to 10^{-10}s.
The experimental part of the work has been done using time-of-flight spectrometer NEAT at BENSC/Berlin [4]. We have investigated the behaviour of molecular hydrogen in the bulk and confined into cages of different dimensions of ice-based clathrates and used for this purpose the composite THF - H_2O - H_2 and H_2O - H_2 systems. THF stands for the tetrahydrofuran molecule. The structure of the H_2 filled ice-clathrates consist from 16 small cavities in the shape of pentagonal dodecahedron (5^{12}) and 8 large cavities in the shape of hexakaidecahedron ($5^{12}6^4$). In the ternary clathrate the THF molecules take up place in the larger cages, and the molecular hydrogen occupies only small cages of 5.02Å accessible diameter with one molecule / cage. In the fully hydrogenated H_2O - H_2 system hydrogen can occupy both the small and the large cages. The latter ones are of 6.67Å accessible diameter and accommodate two H_2 molecules per cage.

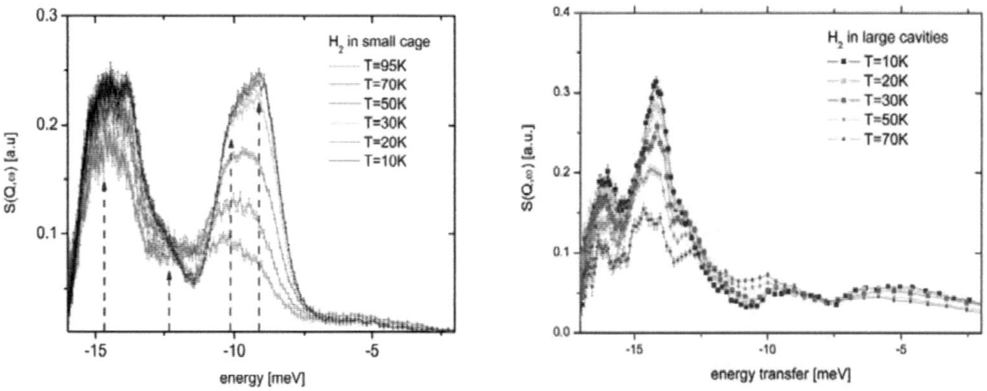

Fig.1 Dynamic structure factor of hydrogen in small (a) and large (b) cages measured by neutron TOF spectroscopy. Dashed lines indicate the position of first rotational transition in the bulk H_2 (blue line) and additional modes which are observed only in the small cage (black lines).

The scattering function of hydrogen molecules is described by discrete spectrum of rotational and vibrational levels [5]. The energies of rotational levels are given by $E_{rot} = J(J+1)B$, where J=1,2,3,... and B=7.35meV is the rotational component and vibrational levels are given by $E_{vib}=(\nu+ \frac{1}{2})\hbar\omega_{vib}$ with vibrational constant ω_{vib}=516 meV. The measurements have been taken at 2 to 100K temperature range. The dynamics response of the bulk hydrogen in the energy range studied by inelastic neutron scattering shows one single peak corresponding to the first rotational transition at 14.7 meV. The response changes once the hydrogen is confined into the cages. The first rotational transition splits into tree components indicating anisotropic cage environment and rather weak, Van-der-Waals type interactions between H_2 and the ice framework. The splitting in the large cage is larger than the splitting in the small cage indicating higher anisotropy of the large cage.

The surprising feature is, however, that in addition to the first rotational transition we observe a new additional dynamic feature in 8-12 meV energy range consisting of several components [fig.1a]. The positions of the peaks depend on the temperature in different fashion: while some components soften with temperature increase, the others exhibit strong anharmonic behavior and move to the higher frequencies. Remarkable is that such dynamic response is absent when the hydrogen is placed into the large cages (fig.1b). Considering that the guest – host cage wall interactions in both cages are of the same nature, it must be the effect of confinement dimensions that gives rise to this additional dynamic feature.

To check whether the translational motion of hydrogen molecules inside the cages contributes to the observed excess of density of states (DOS) we have performed a series of molecular dynamics simulations [4]. For our molecular dynamics simulations we have used the Materials Studio and VASP program suites. The Materials Studio Discover program has been used to perform classical molecular dynamics simulations on hydrogen inside a clathrate hydrate stabilized with tetrahydrofuran, using Compass forcefield and Ewald summation for both Van-der-Waals and Coulomb interactions. We simulated the behaviour of 8 unit cells for times up to 1 ns at various temperatures using a timestep of 1 fs. VASP has been used for ab-initio quantum-mechanical molecular dynamics, where only one primitive cell of the clathrate on the Dirac cluster has been simulated at various temperatures and with different hydrogen loadings, for 4 ps using a timestep of 1 fs. Using the nMoldyn program suite the dynamic structure factor S(Q,ω) has been calculated from Materials Studio and VASP trajectories and compared directly with our experimental results.

The calculated S(Q,ω) [4] shows indeed that in the small cage we observe a translation motion (rattling) in the 8-12 meV energy range. The rattling modes shift to low frequencies and become less pronounced once the dimensions of the cage are increased. These exactly are the tendencies observed in our experiments. It must be mentioned, that the available molecular dynamics tools cannot handle the internal dynamics of the hydrogen molecule, including the rotational motion.

However, the translational motion alone cannot explain the difference between the intensities observed in the small and in the large cages. Considering the amount of hydrogen in the cages and the numbers of cages in the structure, the intensity of the rattling modes should be about the same in both cages. The excess of DOS observed for hydrogen in the small cages must have an additional dynamic origin. At these temperatures, beyond the translational motion, only the molecular rotation is excited and can contribute to the S (Q,ω). The energies of rotational and translation dynamics are pretty close, and the coupling between these two modes can lead to new, hybrid type motion. On the one hand, this can explain the observed excess DOS. On the other hand, the onset of the hybrid motion reflects strong coupling between modes and thus is expected to decrease the kinetic energy of the confined molecular hydrogen. This in turn increases its effective "attachment" to the framework (cage walls) and thus increases the temperature-pressure stability of the entire H_2 clathrate system.

References

[1] G.A. Jeffrey, in *Inclusion Compounds*, eds. J.L. Atwood, J.E.D. Davies, & D.D. MacNicol, Academic, New York, 1984.
[2] W.L. Mao, H.K. Mao, A.F. Goncharov, V.V. Struzhkin, Q.Hu Guo, J. Shu, R.J. Hemley,M. Somayazulu and Y. Zhao, Sceince 297 (2002) pp.
[3] H. Lee; J.-W. Lee, D.Y. Kim, J. Park, Y.-T. Seo, H. Zeng, I.L. Moudrakovski, Nature **434**, 743 (2005)
[4] M. Russina, E. Kemner, M. Celli, L. Ullivi, F.Mezei to be published
[5] J.U.Young and J.A. Koppel, Phys.Rev.135A,603 (1964)

AUTHOR INDEX

Institute of Physics
Dirac House, Temple Back
Bristol BS1 6BE UK